EAI/Springer Innovations in Communication and Computing

Series editor

Imrich Chlamtac, European Alliance for Innovation, Gent, Belgium

Editor's Note

The impact of information technologies is creating a new world yet not fully understood. The extent and speed of economic, life style and social changes already perceived in everyday life is hard to estimate without understanding the technological driving forces behind it. This series presents contributed volumes featuring the latest research and development in the various information engineering technologies that play a key role in this process.

The range of topics, focusing primarily on communications and computing engineering include, but are not limited to, wireless networks; mobile communication; design and learning; gaming; interaction; e-health and pervasive healthcare; energy management; smart grids; cognitive radio networks; computation; cloud computing; ubiquitous connectivity, and in mode general smart living, smart cities, Internet of Things and more. The series publishes a combination of expanded papers selected from hosted and sponsored European Alliance for Innovation (EAI) conferences that present cutting edge, global research as well as provide new perspectives on traditional related engineering fields. This content, complemented with open calls for contribution of book titles and individual chapters, together maintain Springer's and EAI's high standards of academic excellence. The audience for the books consists of researchers, industry professionals, advanced level students as well as practitioners in related fields of activity include information and communication specialists, security experts, economists, urban planners, doctors, and in general representatives in all those walks of life affected ad contributing to the information revolution.

About EAI

EAI is a grassroots member organization initiated through cooperation between businesses, public, private and government organizations to address the global challenges of Europe's future competitiveness and link the European Research community with its counterparts around the globe. EAI reaches out to hundreds of thousands of individual subscribers on all continents and collaborates with an institutional member base including Fortune 500 companies, government organizations, and educational institutions, provide a free research and innovation platform.

Through its open free membership model EAI promotes a new research and innovation culture based on collaboration, connectivity and recognition of excellence by community.

More information about this series at http://www.springer.com/series/15427

Josephina Antoniou

Game Theory, the Internet of Things and 5G Networks

Utilizing Game Theoretic Models to Characterize Challenging Scenarios

Josephina Antoniou
University of Central Lancashire
Larnaka, Cyprus

ISSN 2522-8595   ISSN 2522-8609 (electronic)
EAI/Springer Innovations in Communication and Computing
ISBN 978-3-030-16846-9   ISBN 978-3-030-16844-5 (eBook)
https://doi.org/10.1007/978-3-030-16844-5

This Springer imprint is published by the registered company Springer Nature Switzerland AG.
The registered company address is: Gewerbestrasse 11, 6330 Cham, Switzerland

Preface

With the recent deployment of fourth generation technology (4G), the fifth generation of mobile communication technology (5G) is slowly emerging to support the Internet of Things (IoT), where millions of sensors and mobile devices will be deployed in order to provide data for smart homes, smart buildings and smart cities. 5G networks will have to handle data (collection, storage, mining, analysis, etc.) gathered from a very diverse set of sources like traffic, weather, security incidents, and crowds, and additional issues surfacing due to IoT deployment and the increasing use of wearable devices. Since any communication network, such as IoT, is a multi-entity system, decisions are taken by different system entities. All entities are motivated to make decisions that optimize their own potential benefit, whether this is experience, profit, resource usage or any other factor that may result in high utility measurements or high satisfaction for these entities.

The strengthening of the user role allows a user of the 5G communication network to decide whether or not to participate in a network and offer value to that network. For example, a user may select it to provide a requested service, its criterion being user satisfaction in terms of, for example, experience (e.g. service quality) and cost. Then user experience can be improved by the consideration of context information, i.e. location, device capabilities, environment, preferences, etc. Alternatively, the network operator may decide which of the users to admit, how to allocate available resources to the participating users and how to design certain mechanisms in order to maximize its capacity. All these decisions are driven mainly by the network's revenue maximization criterion.

In this book, we utilize Game Theory in order to model, analyse and finally propose solutions (in terms of strategies for the involved entities) for a number of representative types of interactions in 5G communication networks. The interacting entities are the mobile user devices, IoT nodes and vehicular nodes participating in the 5G network, on the one hand, and the actual network operators and service operators, on the other hand. Game Theory has been extensively used in networking research as a theoretical decision-making framework. Game Theory provides appropriate models and tools to handle multiple, interacting entities attempting to make a decision and seek a solution state that maximizes each entity's utility.

To this end, in this book four selected interactive situations have been analysed using existing game theoretical models, and theoretical conclusions are drawn for each interactive situation; the theoretical conclusions are further reinforced with appropriate numerical results in several cases. Summarizing the work presented, we begin with a scenario where multiple IoT nodes cooperate to achieve the adoption of a password generation mechanism to protect them against malware attacks. Next, we turn our attention to the offloading scenario where mobile nodes take on the role of content repeaters to provide an alternative communication path through the network, subsequently increasing the network coverage. Therefore, 5G communication service is enabled by data offloading from infrastructure transmission points onto the mobile users that act as content repeaters, for a monetary payoff. The third scenario explores the interaction between a receiving node and a sending node in a vehicular network, in order to motivate trust between them. The final scenario deals with the bargaining situation between two provider entities, namely a service provider and a cloud provider, attempting to partition a service payment optimally.

Last but not least, we would like to thank the kind support from the EAI/Springer Innovations in Communications and Computing managing director and this book editor, Ms. Eliska Vlckova. We also appreciate the hard work of all those who have worked together to push forward the publication of the book.

Larnaka, Cyprus   Josephina Antoniou

Contents

Chapter 1
Game Theory and Networking

Abstract With the 4th generation technology (4G), only being deployed for a few years, 5G technology is slowly emerging to support the Internet of things (IoT), where millions of sensors and mobile devices will be deployed in order to provide data for smart homes, smart buildings and smart cities. 5G networks will have to handle data (collection, storage, mining, analysis, etc.) gathered from a very diverse set of sources like traffic, weather, security incidents, crowds, etc. Data analytics and network management are thus necessary for 5G deployment. IoT includes sensors and mobile devices that gather data and perform data mining on data to anticipate certain circumstances, including human behaviour. Some issues that may arise from the deployment of 5G and IoT include new security issues because of the IoT deployment, and additional issues surfacing due to the increasing use of wearable devices. Since any communication network, such as IoT, is a multi-entity system, decisions are taken by different system entities. Such decision-making entities are the "things" in IoT, i.e. the sensors comprising the sensor networks that offer the capability to create smart spaces and applications, the users of the IoT, the content and service providers using the IoT as their infrastructure, etc. All entities are motivated to make decisions that maximize their own potential benefit, whether this is experience, profit, minimal resource usage or any other factor that may result in high utility measurements or high *satisfaction* for these entities. The book will explore specific interactions using a game theoretic framework and offer the equilibriums that maximize the payoffs of the interacting entities.

Keywords 5G networks · Game theory · Emerging technologies · Networking · Internet of things

1.1 Introducing 5G Networks

With the 4th generation technology (4G), only being deployed for a few years, 5G technology is slowly emerging to support the Internet of things (IoT), where millions of sensors and mobile devices will be deployed in order to provide data

© Springer Nature Switzerland AG 2020
J. Antoniou, *Game Theory, the Internet of Things and 5G Networks*,
EAI/Springer Innovations in Communication and Computing,
https://doi.org/10.1007/978-3-030-16844-5_1

for smart homes, smart buildings and smart cities. 5G networks will have to handle data (collection, storage, mining, analysis, etc.) gathered from a very diverse set of sources like traffic, weather, security incidents, crowds, etc. Data analytics and network management are thus necessary for 5G deployment. IoT includes sensors and mobile devices that gather data and perform data mining on data to anticipate certain circumstances, including human behaviour.

Some issues that may arise from the deployment of 5G and IoT include new security issues because of the IoT deployment, and additional issues surfacing due to the increasing use of wearable devices. Regarding security in IoT, we need to consider that if end nodes in the IoT are compromised, large botnets or malicious robot networks are created that use the combined power of the compromised IoT nodes to launch attacks. These can include the mobile devices of innocent users that are geographically spread or collocated. Regarding wearable devices, the issue to consider is when such devices are used to monitor a user's heart rate or calories burnt, i.e. are used to collect personal and sensitive data, and are connected wirelessly to their mobile device, even using the Internet to launch a relevant application, issues arise such as private data being sold to interested parties (maliciously or through agreements of the service providers with specific companies, e.g. advertising companies).

Since any communication network, such as IoT, is a multi-entity system, decisions are taken by different system entities. Such decision-making entities are the "things" in IoT, i.e. the sensors comprising the sensor networks that offer the capability to create smart spaces and applications, the users of the IoT, the content and service providers using the IoT as their infrastructure, etc. All entities are motivated to make decisions that maximize their own potential benefit, whether this is experience, profit, minimal resource usage or any other factor that may result in high utility measurements or high *satisfaction* for these entities. Therefore it is important to decide how such entities in different situations can in fact maximize their own potential impact or their utility measurements.

The book will explore specific interactions using a game theoretic framework and offer the equilibriums that maximize the payoffs of the interacting entities. Implementations of specific numerical models will be presented to support the model evaluations.

5G networks are the evolution of the mobile network technology. Mobile and wireless communication networks comprise a mature industry with increasing user penetration, which has been globally available for some time and is usually deployed as one or another technology, often in competition with each other. 5G communication networks are envisioned to be based upon a common, flexible and scalable convergence platform, where different networking technologies, terminals and services can coexist. Furthermore, and especially with the increasing deployment and use of cloud infrastructures, network access is being decoupled from service provision, resulting in an increasing collection of new services offered by the newly introduced service providers. Decoupling of network access has favoured the idea of convergence, i.e. the co-existence and integration of the various networking technologies in order to act as a unified platform combining their resources to best serve the increasing user requirements.

Such networking technologies may employ different wireless and mobile technologies, and will be interconnected by an IP packet-switched core network. This IP core network deals with all network functionality and coordinates the participating networks, in order for the system to behave as a unified platform. The IP multimedia subsystem (IMS) in past 3G and 4G networks has been an architectural subsystem for the control and provisioning of IP multimedia services over a packet-based core. The IMS was first standardized by the 3rd Generation Partnership Project but it has since been adopted and contributed to by other standardization bodies, including ITU-T and ETSI. The IMS promises a scalable, integrated platform that enables the creation of new services and facilitates the convergence of telecommunications and Internet services. The IMS also promises to achieve fixed mobile convergence by enabling the seamless distribution of services over fixed and mobile broadband networks. However, this requires that IMS services need to be able to cater for heterogeneous access networks and varying user end terminal capabilities. This adaptation ability presents a significant challenge for multimedia services as such adaptation is no easy task.

Moreover, the co-existence of different networking technologies, terminals and services brings forth a new communication paradigm, employed in its totality in 4G, which is *user-centric* [16], i.e. the user is no longer bound to only one access network but may indirectly *select* the *best* available access network(s) to support a service session. Upon a new service request or a particular traffic demand, as well as any dynamic change affecting the session, e.g. mobility, one (or a group) of the participating access networks needs to be selected in order to support the session. Therefore, converged communication networks need to be equipped with a network selection mechanism to assign the *best* access network(s) to handle service activation, or any dynamic session change. Such decision may result in the selection of a single access network or even a group of networks in 4G networking platforms.

A 5G network, as considered in this book, is a multi-entity system, i.e. decisions are taken by different system entities. The decisions serve to provide the means to efficiently support a requested service, with most likely much higher demands than 3G and 4G services, to the appropriate user. At least, two entities of decision-making are necessary; one representing what the user needs and another representing what the network can provide. Thus, the minimum number of entities responsible for these decisions are: (i) the user (requesting services) and (ii) the network operator (responsible to support the user requested services). All decision-making entities can be driven by satisfaction-maximization functions, which are based on the individual entity's criteria [10]. Nevertheless, the roles can easily change in a 5G network with the user often acting as an infrastructure component and the network operator interacting with users or other service and content providers on a peer-to-peer basis.

The strengthening of the user role allows a user of the 5G communication network to decide whether or not to participate in a network and offer value to that network, selecting it to provide a requested service, its criterion being *user satisfaction* [17] in terms of, for example, experience (e.g. service quality) and cost; user experience can be improved by the consideration of context information, i.e. location, device capabilities, environment, preferences, etc. On the other hand, the network operator may decide which of the users to admit, how to allocate available resources to the participating users and how to design certain resource management mechanisms in order to maximize its capacity. All these decisions are driven mainly by only one criterion: the network's revenue maximization.

Therefore, considering the example entities of the user and the network we may initially assume the following: on the one hand, the network operator sets the *price per user* for using network resources based on its own revenue-maximization criterion and, on the other hand, the user, given the price set by the network as well as additional parameters that comprise its own satisfaction function, makes the decision whether or not to participate in the particular network, always subject to the network's own admission policy [11, 13].

1.2 Datagram Networks and Technology Convergence

Traditionally, communication networks provide one of the two different types of network services: circuit-switched service and datagram. The typical model for a circuit-switched network is the public switched telephone network (PSTN). Once a call is established through the various switching centres, the path from source to destination is also established like a circuit. The transport characteristics in a circuit-switched path may be predetermined and the connection is said to be predictable in terms of the QoS it can deliver. On the other hand, the typical model for a datagram type network is the Internet, where data is broken into small units, called packets, for transmission through the network using only source and destination addresses and trusting that existing routing technologies will deliver each packet to the correct destination; this type of connection is referred to as packet switched and it is not easy to predict the QoS it can deliver.

Voice communication networks have hitherto embraced the circuit-switched approach because of the strict delay constraints of the voice service, while data networks have been successful in using the datagram or packet-switched approach since delay bounds are more relaxed. However, the rapid deployment of emerging information technologies, including mobile communications and Internet-related technologies in general, plus the widespread acceptance of the Internet protocol (IP) been some of the driving factors towards integrating voice and data networks of different technologies into a unified infrastructure, introducing the idea of convergence, i.e. combining voice, data and other media such as video, audio and fax, into what appears to be a single network interface but may comprise the integration of heterogeneous wired and wireless networks.

Despite the inability of IP to offer guaranteed QoS, IP-based networks upon deployment became a very popular and cost-efficient communication solution. Even though, there were no QoS guarantees from the IP itself, additional QoS mechanisms improved QoS-provisioning so that a multitude of services could be successfully deployed over IP. The IP trend evolved to the all-IP paradigm, i.e. the IP supporting next generation communication networks, which are characterized by the convergence of heterogeneous access technologies, by deploying an IP-based core where the participating heterogeneous access technologies converge. Consequently, we may define convergence to be the creation of an environment that can ultimately provide seamless and high-quality broadband mobile communication service and ubiquitous service through wired and wireless integrated networks without spatial and temporal constraints by means of connectivity for anybody and anything, anytime and anywhere.

A converged network is a datagram type network in principle, combining traditionally circuit-switched and datagram networking into a single packet-based network. However, it needs to address reliability in order to be successful. The IP, which was designed to support datagram type networks, is not sufficient in itself to transmit real-time traffic such as voice and video. Additional protocols are needed. Many organizations are addressing this challenge, such as ETSI with TIPHON (telecommunications and internet protocol harmonization over network), ITU-T with H.225, and H.245 standards for multimedia communication, and IETF with SIP (session initiation protocol), SDP (session description protocol), RTP (real-time protocol), RSVP (resource reservation protocol) and others that facilitate real-time communications over IP-based networks.

A user participating in a converged network enjoys several advantages. Although there is a multitude of underlying technologies, only one interface is perceived by the user, which means the user does not need to know how to manage multiple systems and in terms of support the user contacts only one organization, the converged platform administrator, for an increased number of available services. The access networks themselves enjoy increased user participation and economic benefits through possible cooperation with other participating networks to better support the converged network services. In addition, network performance becomes more efficient since the platform administrator can enforce certain policies to all the participating networks and due to access network synergies, the participating networks are expected to become more competitive.

In terms of challenges to overcome, the ones that a converged infrastructure needs to deal with are threefold: management challenges, architectural challenges and quality challenges (i.e. service quality perceived by the user). In terms of management, the appropriate mechanisms must consider the heterogeneity of the network and perhaps the different ownership, and in turn satisfy several challenging demands, as, for example, high throughput, authorization and billing, security, fairness, deployment costs and interoperability issues, ensuring that (a) environment heterogeneity is transparent to the users who simultaneously have maximum choice in their selection of and access to a network with suitable QoS and information services and (b) that information providers can reach all users. In terms of architecture, the functionalities and interactions of additional components that will enable the convergence are necessary to ensure the efficient operation of the system and allow for the management mechanisms to resolve the issues mentioned above. Management and architecture issues have various common areas in terms of design and implementation. Quality, and particularly improving user experience, touches upon network and resource management and moves on to areas such as modifications to user devices, improvements to applications and services, enhancing user capabilities (e.g. exploring context-awareness), etc.

Since network access is being decoupled from service provision, the result is an increasing collection of new services offered by the newly introduced service providers. The user-centric view evolving from this advancement introduces a setting where the user can choose or take active part in the selection of the best available network to serve a requested service, and no longer has to subscribe to one particular access provider. Hence, network selection mechanism is a newly introduced resource management mechanism in a converged network, which handles the selection of the best network to satisfy a service request, aiming to improve both user experience and network motivation to offer its resources (through increased revenues). The significance of this advancement is the empowerment of the role of the mobile user as a part of a communication network that continues similarly in 5G communication networks as well.

Mobile communication networks have already motivated cooperation at the level of networks, since they promote the co-existence of different networks on a convergence platform such that the actual network serving a particular request is transparent to the mobile user. Network selection is the mechanism responsible to select the network(s) to serve a particular request out of the set of available networks. Selecting the network(s) to serve a user demand that satisfies both the user preferences and the network constraints may become a challenging task. Since network selection is a mechanism involving important decision-making, it must take into consideration some strategical planning on behalf of the entities involved in this decision. Given that game theory is a theoretical framework for strategical decision-making, it has been a very popular approach among many of the recently presented research works. In addition to the most commonly seen non-cooperative games [5, 6], several cooperation schemes have emerged to propose solutions in situations where limited resources or need for quality guarantees exists [1, 22].

Furthermore, interconnection of networks results in various forms of infrastructure cooperation, which although socially desirable may not always be beneficial for the participating networks. Therefore, to avoid defection from cooperation, this is sometimes made mandatory for the cooperating networks, since higher data rates, better coverage, improved energy consumption as well as more accurate location estimation could result from infrastructure cooperation [23]. Underlying principles of cooperative techniques as well as several applications demonstrating the use of such techniques in practical systems are demonstrated in [12]. Given such cooperative environments, enhanced services can be designed to take advantage of the existing cooperation between networks and network components. Moreover, the development of wireless and mobile communications industry and technology in the near future depends largely upon the merging of the Internet with the mobile world, requiring the cooperation of the involved heterogeneous network entities, to achieve attractive scenarios for the future mobile user.

1.3   Introducing Game Theory for Decision-Making in Networking Environments

We recognize that in a 5G network there exists the need to consider the *interaction between each user and each one of the available networking technology providers, service providers and content providers* as well as any interactions between the providers themselves, in order to improve decision-making. Interactions between entities with conflicting interests follow action plans designed by each entity in such a way as to achieve a particular selfish goal and are known as strategic interactions. *Game theory* is a theoretical framework that studies strategic interactions, by developing models that prescribe actions in order for the interacting entities to achieve satisfactory gains from the situation.

In this book, we utilize game theory in order to model, analyse and finally propose solutions (in terms of strategies for the involved entities) for a number of representative types of interactions in 5G communication networks. The interacting entities are the *mobile user devices, IoT nodes,* and *vehicular nodes* participating in the 5G network, on the one hand, and the actual *network operators* and *service operators*, on the other hand. Well-designed strategies can enhance capacity planning on a dynamic basis (for example, on a day-to-day), by allowing the constituent technologies and participating nodes and consequently the whole 5G network to make the best use of the available capacity, at any given horizon.

Game theory has been extensively used in networking research as a theoretical decision-making framework. Game theory is mainly based on [9, 14] appeared formally in the 1940s in a text by John von Neumann and Oskar Morgenstern [27], although the ideas of games and equilibria are found as early as 500 AD in the Babylonian Talmud, which is the compilation of ancient law and tradition for the Jewish Religion [3], as well as in the 1800s in Darwin's *The Descent of Man, and Selection in Relation to Sex* [8]. Game theory grew more popular in the 1950s and the 1960s with important contributors such as John Nash [20], Thomas Schelling [25], Robert John Aumann [2] and John Harsanyi [15], as well as in the 1970s with Reinhard Selten [26], giving all the above scientists Nobel Prize awards in Economics in 1994 (John Nash, John Harsanyi and Reinhard Selten) and in 2005 (Robert John Aumann and Thomas Schelling). It is worth mentioning some more recent, important contributors of game theory such as Ariel Rubinstein who contributed in the theory of bargaining [24] and Vincent P. Crawford for his work with redefinition of equilibria [7].

1.3.1   Introducing Utility Theory

Making a decisions is a challenging task because it must consider abstract notions affecting a particular decision, e.g. the preferences of the agents participating in the decision-making. It is often convenient to represent preferences with a utility function and reason indirectly about preferences with utility functions. Utility functions express the desirability of a state for an agent.

Utility theory is the area that studies the quantification of the preferences between different outcomes of various plans. Usually a decision is based upon the outcome of the agents' utility functions and the probabilities of occurrence of the relevant events. Let X be the set of all the mutually exclusive options an agent could prefer. The agent's utility function may be defined as $u : X \rightarrow R$ and it ranks each option in the consumption set X. If $u(x) \leq u(y)$, then the agent strictly prefers or is indifferent to option y in relation to option x. A utility function $u : X \rightarrow R$ rationalizes a preference relation $\preceq$ on X, if for every $x, y \in X, u(x) \leq u(y)$ if and only if $x \preceq y$.

Therefore, a utility function represents a complete, reflexive, transitive and continuous preference relation. Such a function represents an ordinal concept, i.e. it gives an ordering or ranking of preferences but does not reflect the amount of preference, e.g. if $u(x) = 2$ and $u(y) = 6$, it is not implied that y is *three times better* than x. According to the specific utility function, different combinations of input parameters may result in the same utility.

Each possible output of a utility function may be referred to as a utility level and the combinations of input parameters that result into a specific utility level may be drawn into a curve, known as an indifference curve because all points on an indifference curve have equal preference. The more sets of input parameters we consider, the larger the collection of indifference curves that will be created and consequently the description of the user's preferences will improve. Furthermore, by comparing all possible combinations of input parameters we may create the complete collection of the user's indifference curves, each with its assigned utility level, thus completely representing the user's preferences. The collection of all indifference curves for a given preference relation is an indifference map and it is equivalent to a utility function.

1.4   Game Theory as an Analytical Tool

This book considers the network selection mechanism as a decision resulting from the interaction between the user and the access networks available to the user. Describing and analysing entity interactions is a situation that makes a good candidate to be modelled using the theoretical framework of game theory. Game theory provides appropriate models and tools to handle multiple, interacting entities attempting to make decision and seeking a solution state that maximizes each entity's utility. Game theory has been extensively used in networking research as a theoretical decision-making framework, i.e. for routing, congestion control, resource sharing, etc. The network selection decision may benefit from such a theoretical framework that considers decision-making, interacting entities.

Game theory is a theoretical framework that attempts to mathematically capture both human and non-human (computer, animal and plant) behaviour during a strategic situation. A strategic situation is a situation that involves the interaction of two or more entities in which the individual's success depends on the choices of others attempting to find equilibria between the entities (called the *players*), i.e. sets of strategies (action sequences) that players will unlikely want to change. Therefore, game theory can be used to model situations of interaction and offer solutions so that mutually agreeable results can be reached; game theoretic models make the assumption that the entities make rational choices, i.e. choices that are profitable according to each entity's own interpretation of profit.

In order for a strategic situation to become a *game* between two or more players, there must be a mutual awareness of the participants regarding the cross-effect of their actions. A strategic situation, where the actions of a participant may alter another's outcome, is primarily characterized by the players' strategies. In addition, a strategic situation contains other elements that must be taken into consideration when modelling such a situation as a game, e.g. chance and skill (elements that are not easily controlled or modified). The idea of *signals*, i.e. specific observations, is often used to reach conclusions regarding a situation's or a player's information that would be unclear otherwise. For example, signals may help a player to recognize if his opponent is exaggerating or even lying, since signals offer objective evidence.

Game models, i.e. models of specific strategic situations, may be categorized in various ways due to the several elements that they contain. A usual categorization is made by looking at the players' movements; if they are sequential we have an extensive game form, whereas if they are simultaneous the game form is referred to as normal. Furthermore, an interaction may happen only once or repeatedly; in the first situation we are faced with one-shot game models, while the second situation requires repeated game models. An additional dichotomy is whether the players are in complete conflict, where the model employed is a non-cooperative one, or they have some commonality, where a more cooperative game model may be more appropriate. Finally, another important categorization is whether we are dealing with a game where the players have complete information about all actions taken or only partial information. When characterizing a game it is important to keep in mind the various possible categorizations of game models in order to better describe the required strategic situation as completely as possible.

As mentioned above, strategic games employ the element of rationality. Usually rationality may imply that players are perfect calculators and flawless followers of their best strategies, although this may be defined not to be the case. Therefore, rationality may be better described to be the players' knowledge of their own interests based on each player's own value system. Based on this element of rationality the players calculate their possible strategies. Depending on whether the game is normal or extensive strategies may consist of single actions or sequences of actions and each strategy gives a complete plan of action, considering also reactions to actions that may be taken by the opponent. Strategies may be pure, i.e. provide complete definitions of how a player will play in the game (his moves), or mixed, i.e. assignments of a probability to each pure strategy.

Games are motivated by *profitable* outcomes that await the players once the actions are taken. These outcomes are referred to as payoffs. Payoffs for a particular player capture everything in the outcomes that the particular player cares about. If a player faces a random prospect of outcomes, then the number associated with this prospect is the average of the payoffs associated with each component outcome, weighted by their probabilities.

The solution to a strategic game is derived by establishing equilibria. Equilibria may be reached during the interaction of players' strategies when each player is using the strategy that is the best response to the strategies of the other players. The idea of equilibrium is a useful descriptive tool and furthermore, an effective organizing concept for analysing a game theoretic model. For simultaneous-move games the Nash equilibrium is used as a solution concept, where every player's action is the best response to the actions of all the others. For sequential-move games, also used in this book, the equilibrium used is known as the subgame perfect equilibrium or the rollback equilibrium and is generated by the method of backward induction, which captures the concept of looking ahead and reasoning back. This is because in such games the players must use a particular type of interactive thinking; players plan their current moves based on future consequences considering also opponents' moves. Therefore, the equilibrium in such a game must satisfy this kind of interactive thinking and backward induction does exactly that, by planning the best responses backward starting from the last moves and moving to the first.

Moreover, an easy way to visualize a sequential-move game is by illustrating the game using tree diagrams made from nodes and branches (a.k.a. game trees); they are joined decision trees for all of the players in the game, illustrating all of the possible actions that can be taken by all of the players and indicating all of the possible outcomes of the game. This book employs the notion of game trees.

1.5   A Glance at Well-Known Game Theoretic Models

The following subsections elaborate on certain game theoretic aspects that are used in this book. Section 1.5.1 presents and discusses the single-shot as well as the iterated *prisoner's dilemma* game, which is the most well-known game model between two interacting entities. Section 1.5.2 provides an overview of bargaining games and focuses on two well-known two-player models, the *Nash bargaining game* and the *Rubinstein bargaining game*. In Sect. 1.5.3 we focus on games of incomplete information, whereas Sect. 1.5.4 presents the repeated game, which models iterative interactions. Finally, Sect. 1.5.5 describes the interaction of coalitions of players instead of single players. Note that all these models, which we further exploit in this book, employ the element of cooperation to provide solutions to the various situations. This is appropriate in our study, where we aim to motivate cooperation of the various heterogeneous entities, so as to achieve satisfactory outcomes for all entities involved.

1.5.1   The Prisoner's Dilemma

The *prisoner's dilemma* and *Iterated prisoner's dilemma*[1] have been a rich source of research material since the 1950s. However, the publication of Axelrod's book in 1984 [4] was the main driver that this research was brought to the attention of other areas outside of game theory, as a model for promoting cooperation.

The prisoner's dilemma is basically a model of a game, where two players must decide whether to cooperate with their opponent or whether to defect from cooperation. Both players make a decision without knowing the decision of their opponent, and only after the individual decisions are made, these are revealed. The story behind this model is the following: *Two suspects are arrested by the police. Since the policemen have insufficient evidence for their conviction, they separate the two suspects and offer them a deal. If one testifies (i.e. defects from the other) against the other and the other remains silent (i.e. cooperates with the other), the betrayer goes free and the silent goes to jail for 10 years. If both remain silent, then they both go to jail for only 6 months with a minor charge. If both testify, each gets a 5-year sentence.* Each suspect must choose whether to testify or to remain silent, given that neither learns about the decision of the other until the end of the investigation. What should they do?

Mutual cooperation has a reward for both of receiving the punishment which is the least harsh. However, such decision entails the risk that in case the other player defects, then the cooperative player will receive the harshest punishment of 10 years. Given the risk of cooperation, it is very tempting to defect because if the opponent cooperates, then defecting will result in the best payoff, which is to go free, although, if the other opponent also defects, then an average punishment will be received by both (i.e. 5 years). The decision of what to do comes from the following reasoning: *If a player believes that his opponent will cooperate, then the best option is certainly to defect since the payoff will be to go free. If a player believes that his opponent will defect, then by cooperating he takes the risk of receiving the harshest punishment, i.e. 10 years, thus the best option is again to defect and share the average punishment of 5 years with his opponent.* Therefore, based on this reasoning, each player will defect because it is the best option no matter what the opponent chooses. However, this is not the best possible outcome of the game. The best solution would be for both players to cooperate and receive the least harsh punishment of 6 months.

The desirable cooperative behaviour must be somehow motivated so that the players' selfish but rational reasoning results in the cooperative decision. In fact, what we have described is a *one-shot* prisoner's dilemma, i.e. the players have to decide only once—no previous or future interaction of the two players affects this decision. Cooperation may evolve, however, from playing the game repeatedly, against the same opponent. This is referred to as iterated prisoner's dilemma,

---

[1]The text in Sect. 1.5.1 is based on [18].

which is based on a repeated game model with an unknown or infinite number of repetitions. The decisions at such games, which are taken at each repetition of the game, are affected by past actions and future expectations, resulting in strategies that motivate cooperation. The way motivation may be encouraged in such games will be analysed within the work presented in this book. The repeated game model is briefly overviewed in Sect. 1.5.4.

1.5.2   Overview of Bargaining Games

With the exception of the groundbreaking contributions of John F. Nash [20], bargaining theory[2] basically evolved from the seminal paper by Ariel Rubinstein [24], which made the procedure of bargaining quite attractive, mainly due to the proposed model's simplicity and ease of understanding.

A bargaining situation is an exchange situation, in which two individuals have a common interest to *trade* but simultaneously have conflicting interests about the *price* at which to trade, because the seller would like to trade at a higher price, while the buyer would like to trade at a lower price. Therefore, in a bargaining situation, the two players have a common interest to cooperate but have conflicting interests about exactly how to cooperate. On the one hand, each player would like to reach an agreement rather than disagree; on the other hand, each player wants to reach an agreement that is as favourable to that player as possible.

Therefore, a bargaining situation may be easily seen as a game situation since the outcome of bargaining depends on both players' bargaining strategies, i.e. whether or not an agreement is reached and the terms of that agreement (if one is reached) depend on both players' actions during the bargaining process. Next we provide brief descriptions of the two well-known game models of a bargaining situation between two players, the Nash bargaining game model and the Rubinstein bargaining game model, since we will make use of these models in particular formulations presented in this book.

The Nash bargaining game model defines a solution (known as the Nash bargaining solution) by a fairly simple formula and it is applicable to a large class of bargaining situations. The Nash bargaining example is the situation where two individuals bargain over the partition of a cake of fixed size. Since the cake will be partitioned to the two players, the addition of their partitions should equal the total cake; therefore, the set of non-zero partitions, which sum to the total amount, is the set of possible agreements in the bargaining situation. In the case of disagreement, each player receives a penalty, which is defined according to the bargaining situation under consideration; the definition of a penalty comes from the fact that in bargaining situations the desirable outcome is agreement; thus, disagreement results in a non-satisfactory payoff for the two players. The penalties

---

[2]The text in Sect. 1.5.2 is based on [19].

of the two players are defined as the *disagreement point* of the game. Thus, the *useful payoff* for each player may be defined to be the player's payoff received from the received partition in case of agreement, minus the penalty that would be received in case of disagreement as defined in the *disagreement point*. The unique solution of the cake partition is therefore the unique pair of partitions that maximizes the product of the players' *useful payoffs* and is referred to as the *Nash product* or the *Nash bargaining solution*.

The Rubinstein bargaining game is modelled as a sequential-moves game, in which the players take turns to make offers to each other until agreement is secured. This model has much intuitive appeal, because a lot of real-life negotiations are based on the idea of making offers and counter-offers. From the sequential-moves model of the bargaining process, it is easy to see that if the two negotiating players do not incur any costs or penalties for delaying the agreement decision, the solution to the bargaining game may be indeterminate, because the two players could continue to negotiate forever. Given that there is a cost to each player for delaying, then each player's bargaining power is determined by the magnitude of this cost. Consider for the Rubinstein model that, similarly to the Nash bargaining game model described above, the two players bargain over the partition of a cake of fixed size. The first player proposes a partition; if the second player accepts, agreement is reached and the game is over; otherwise, the second player proposes a different partition, and the process of alternating offers continues until an offer is accepted. However, for each additional negotiation round there is a cost to each player, i.e. the size of the cake becomes smaller. The factor by which the cake gets smaller may be different for each player and it is referred to as the player's discount factor. The Rubinstein bargaining game model has a unique subgame perfect equilibrium, which makes use of the fact that any offer made now by a player should be equal or greater to the discounted best value that the opponent can get in the next period.

Another element that must be considered in bargaining situations is truthfulness of the players. In order to motivate the two bargaining players to be truthful about own information that may affect the bargaining process, there must exist a mechanism that can penalize a player who turns out to lie on its real cost, assuming that it is detectable whether a player has lied or not. Such mechanisms exist and are called *pricing mechanisms* [21], constituting an interesting and very promising way to guarantee truthfulness of the participating networks. In the worldwide literature there is a whole research field that is focused on the development, limitations and capabilities of such pricing mechanisms, the algorithmic mechanism design [21]. Successful paradigms in this context include (combinatorial) auctions and task scheduling using techniques such as the revelation principle [21], incentive-compatibility [21], direct-revelation [21] and Vickrey–Clarke–Groves mechanisms. The algorithmic tools and theoretical knowledge that have already developed in the field of algorithmic mechanism design constitute a fruitful pool for extracting algorithmic tools for enforcing players to truthfulness, through pricing mechanisms, once these tools are customized and further developed for handling the needs of any specific scenario.

1.5.3   The Bayesian Game Model

In several situations where interactions occur, the interacting entities may not have complete information about each other's characteristics. The model of a Bayesian game is designed to model such situations. A player's uncertainty about the opponent's characteristics is modelled by introducing a set of possible states, i.e. probable sets of characteristics that a player may have, also known as a player's *types*. Each player assigns a probability of occurrence to each of the opponent's possible *types*. Therefore, a definition of a Bayesian game is similar to the definition of a normal form game, with the additional elements of the *types* for each player and the corresponding probability of occurrence, as believed by the player's opponent(s).

A Bayesian game can be modelled by introducing nature as a player in the game.[3] Nature assigns a random variable to each player, which could take values of types for each player and associate probabilities with these types. In the course of the game, nature randomly chooses a type for each player according to the probability distribution across each player's type space. The type of a player determines the player's payoff and the fact that a Bayesian game is one of incomplete information means that at least one player is unsure of the type and thus the payoff of another player.

In any given play of a Bayesian game, each player knows his type and does not need to plan what to do in the hypothetical event that he is of some other type. However, when determining a player's best action, he must consider what the other player(s) would do if any of the other possible types were to occur, since any player may be imperfectly informed about the current state of the game. Therefore, a Nash equilibrium of a Bayesian game is the Nash equilibrium of the normal form game in which the set of players includes all possible types for each player, and consequently the set of actions includes all possible actions for each such state of every player considered. In brief, to reach a Nash equilibrium in a Bayesian game, each player must choose the best action available to him, given his belief about the occurrence of his opponent's types, the state of the game and the possible actions of his opponent.

1.5.4   The Repeated Game Model

The model of a repeated game is designed to examine the logic of long-term interaction. It captures the idea that a player will take into account the effect of his current behaviour on the other player's future behaviour. Repeated game models aim to explain phenomena like cooperation, threats and punishment.

---

[3]This approach was proposed by John C. Harsanyi in [15].

The repeated game models offer insight into the structure of behaviour when individuals interact repeatedly, a structure that may be interpreted in terms of social norm. The results show that the social norm needed to sustain mutually desirable outcomes involves each player threatening to *punish* any other player whose behaviour is undesirable. Each player uses *threats* to warn the opponent that such punishment may follow if the threats are credible and if there is sufficient incentive for the player to carry out his threat. Thus, punishment depends on how players value their future payoffs and it may be as harsh as lasting forever, or as mild as lasting for only one iteration.

The model of a repeated game has two kinds: the horizon may be *finite*, i.e. it is known in how many periods the game ends, or *infinite*, i.e. the number of game periods is unknown. The results in the two kinds of games are different, for instance, analysing a finite version of the prisoner's dilemma ends in the conclusion that the players are motivated to cheat as in the one-shot prisoner's dilemma, whereas an infinite version of the prisoner's dilemma results in a motivation for both players to cooperate.

The iterated prisoner's dilemma is a quite popular repeated game model which demonstrates how cooperation can be motivated by repetition (in the case the number of periods is unknown), whereas in the one-shot prisoner's dilemma as well as in the finite version of the Iterated prisoner's dilemma, the two players are motivated to cheat. The main idea is that if the game is played repeatedly, then the mutually desirable cooperative outcome is stable because any deviation will end the cooperation, resulting in a subsequent loss for the deviating players that outweighs the payoff from the finite horizon game (horizon of one or more periods). This emerging stability can be used to explain how in finite populations natural selection (occurring over an unknown number of iterations, thus modelled as an infinite horizon game) favours cooperation when starting from an individual using that strategy and further, the analysis of this phenomenon leads to natural conditions for evolutionary stability in finite populations. Thus, when applying the model of a repeated game to a specific situation or problem, e.g. natural selection, we must first determine whether a finite or infinite horizon is appropriate, based on the characteristics of the realistic situation.

1.5.5   Games of Coalitions

Coalitional games deal with the situation, in which interactions occur between groups of players (coalitions) and thus actions are assigned to coalitions even though individual entities may consider their own preferences, especially when selecting a particular coalition in which to participate. Therefore, a coalitional model is characterized by its focus on what groups of players can achieve rather than on what individual players can achieve.

Furthermore, in coalitional games, the way a coalition operates internally, i.e. among its members, is not considered as important for a coalitional game so that the outcome does not depend on such details. A solution concept for coalitional games assigns to each game a set of outcomes, capturing consequences for the participating coalitions. The solution defines a set of arrangements that are stable in some sense, i.e. that the outcomes are immune to deviations of any sort by groups of players.

In order to determine the solution to a coalitional game, we must first define the way payoffs are assigned to the various coalitions; such assignment can occur per group as a whole, or per group using a particular division arrangement within the group for its members. When payoffs are assigned per group, the players that participate in the same group are associated with the group's payoff and it is not defined how this payoff may be further partitioned among its members. This case of payoff assignment is referred to as transferrable payoff coalitional game. The alternative is known as non-transferable payoff coalitional game, and in such model there exists a rule on how group payoffs are divided among participating players.

A well-known solution concept for a coalitional game model is the *core*. The core is a solution concept that requires that no set of players be able to break away and take a joint action that makes all of them better off. Overall, the idea of the *core* is analogous to that behind a Nash equilibrium of a non-cooperative game, i.e. an outcome is stable if no deviation is profitable. In the case of the *core*, an outcome is stable if no coalition can deviate and obtain an outcome better off for all its members.

1.6   Exploring This Book

Interactions in 5G communication networks become a challenging task due to the heterogeneity of user(s) and the mobile network provider(s), as well as additional end nodes and provider entities, resulting in different and often conflicting interests for these entities. Since cooperation between these entities, if achieved, is expected to be beneficial, we have posed the following question, as a common theme for he models explored in this book: Can cooperation be motivated in interactive situations in 5G networks, and if yes, is it beneficial for the interacting entities?

We have utilized, game theory, a theoretical framework suitable for generating profitable behaviours/strategies for interacting entities in conflicting situations and we have explored its application upon seemingly conflicting interactions occurring in 5G communication networks. In particular, four selected interactive situations have been analysed using existing game theoretical models, and theoretical conclusions are drawn for each interactive situation; the theoretical conclusions are further reinforced with appropriate numerical results in several cases. Overall, these conclusions show that cooperation can indeed be motivated in the selected interactive situations and furthermore, that this cooperation is beneficial for the interacting entities.

This book has listed and discussed how several networking situations have shown that cooperation is a beneficial solution both in node-to-node interactive situations, node-to-network interactive scenarios, as well as in provider-to-provider interactions. In modelling these situations, we use game theory as the theoretical framework through the use of which cooperation can be motivated. Overall, game theory in general and cooperation in particular have been used to solve various networking problems. In this book we turn our focus on interactions in 5G communication networks as the environment in which the selected scenarios are motivated; however, applicable cooperation scenario can be expected in other networking interactive situations.

Summarizing the work presented, we begin with a scenario where multiple IoT nodes cooperate to achieve the adoption of a password generation mechanism to protect them against malware attacks. The selected situation involves the interaction between multiple IoT nodes when they must cooperate to serve a large resource demand that is best served by more than one nodes (for example, if none of those networks can serve the demand on its own). The coalition formation process between multiple nodes has been modelled as a coalition game, in which individual nodes with insufficient resources form coalitions in order to satisfy service resource demands. It has been shown that the coalition game is equivalent to the well-known *weighted voting game*. This equivalence encourages the use of power indices for payoff allocation, as often used to resolve the weighted voting game.

A comparative study of well-known *power indices* representing payoff schemes is provided for the coalition game. Based on conclusions from this study, the study proposes the use of a selected power index, the *popularity power index* (PPI), which associates the popularity of each network to the number of stable coalitions it participates in.

It is shown that the selected power index achieves fairness, in the sense that it only considers the possible coalitions that would be formed if payoffs were assigned proportionally to the nodes' contributions. An analysis of the coalition formation is provided for both transferable and non-transferable payoffs, in order to determine stable coalitions using the core and inner core concepts. The most appropriate power index for the coalition game, from the existing power indices investigated, is a power index that provides stability under the core concept, known as the *Holler–Packel index* (HPI). The core and inner core equilibrium concepts are further investigated to show that coalitions that would be formed using the selected power index, PPI, to assign payoffs, are only coalitions that would be stable under the inner core concept. Therefore, the PPI provides a simple and fair payoff allocation method that is equivalent to a stable cooperative equilibrium solution [1].

Next, we turn our attention to the offloading scenario where mobile nodes take on the role of content repeaters to provide an alternative communication path through the network, subsequently increasing the network coverage. Therefore, 5G communication service is enabled by data offloading from infrastructure transmission points onto the mobile users that act as content repeaters, for a monetary payoff. The aim is to enhance the communication service performance, especially in terms of traffic delays, which is a requirement in 5G networks.

The offloading algorithm is viewed as an ad hoc service where the path of the traffic through the content repeating devices is selected through a negotiation process between the mobile users and the infrastructure points, as well as between the mobile users themselves. The relevant chapter proposes a game theoretic modelling approach for modelling the strategies of the interacting nodes using a game model that resembles an auction model. Specifically, each player may take one of the two roles during an interaction, either the role of a price setter, a bidder, or the role of price taker, an auctioneer. We assume that all packets are homogeneous in the sense that they do not differ in terms of the payoff to the forwarder; therefore, the decision of whether to forward a traffic packet by a node, or to choose the node that will forward the traffic packet (depending on whether a player is currently in the role of price setter or price taker), does not depend on the traffic packet itself. The decisions are taken sequentially, i.e. the price setters advertise their prices and then the price taker selects one of them to forward the traffic packets to.

The third scenario explores the interaction between a receiving node and a sending node in a vehicular network, in order to motivate trust between them. The cooperative nature of this scenario is motivated by an infinitely repetitive game, and appropriate strategies for both the receiving node and the sending node are evaluated in order to select the ones that achieve strong motivation for the two entities towards cooperating and remaining in cooperation during communication; a new adaptive node strategy and consequently a new game profile are presented in this chapter.

The interaction between the receiving and sending nodes has been modelled using game theoretic tools, in an attempt to model an interaction that motivates cooperation. Firstly, the interaction is modelled as a one-shot game, and it is shown that there exists equivalence of a one-shot game model of node interaction to the one-shot *prisoner's dilemma game*. This is an important finding, since the *prisoner's dilemma* is known to have a cooperative solution under certain conditions, as, for example, a repeated game, motivating us to explore this interaction further towards the direction of cooperation. In fact, it has been shown that motivation of cooperation between the two entities is achieved through a *repeated game* model of the repeated interaction.

Exploring existing and new strategies for the two entities, we show that when the strategies used by the players of the repeated interaction model involve punishment to motivate cooperation, then harsher punishments motivate cooperation more easily. However, since practically a node wouldn't choose to employ the harshest punishment, i.e. leaving the interaction forever, we propose a new strategy that uses an *adaptive* punishment method in the repeated interaction game. The adaptive strategy is inspired by the 5G network model, which assumes that the participating nodes are adaptive entities with *knowledge* about other nodes' behaviours that change over time. For that, an internal state for a participating node as an adaptive entity is defined and used in the game model scheme.

The adaptive strategy motivates cooperation and achieves satisfying results in terms of motivation and in terms of payoffs, becoming the strategy of choice for a node when compared to the other strategies examined in the relevant chapter. As a consequence, it has been shown that a profile of the repeated interaction game where the receiving node employs the adaptive punishment method and the sending node employs the well-known *tit-for-tat* strategy generates the most profitable payoffs for both players.

The final scenario deals with the bargaining situation between two provider entities, namely a service provider and a cloud provider, attempting to partition a service payment optimally. The interaction is modelled and resolved using the Nash bargaining game model, and accordingly the Nash bargaining solution, which is equivalent to the immediate resolution of the widely used Rubinstein bargaining game. Truthfulness is an issue that must always be considered in such bargaining situations, and thus, a Bayesian game model helps to propose a way to induce truthfulness from the participating players in the last chapter.

The specific interaction involves two providers interacting to support a service with additional data protection requirements (e.g. for a health monitoring service). The two providers must cooperate to partition the service payment, since the fact that two providers support the service is transparent to the user offering the service payment. This interaction between two providers has been modelled using game theoretic tools, in such a way as to motivate cooperation in terms of partitioning the available payment for the particular service.

The payment-partition model has been shown to be equivalent to the well-known *Rubinstein bargaining game*, if the agreement in the Rubinstein bargaining game is reached from the first negotiation period. In addition, the chapter shows that there exists equivalence between the payment-partition game and the *Nash bargaining game*, due to the equivalence of the Nash bargaining game to the Rubinstein bargaining game, if the agreement in the Rubinstein bargaining game is reached from the first negotiation period. Thus, an optimal solution for the payment-partition game exists and is based on the *Nash bargaining solution*, resulting in a partition determined by the cost each of the two providers has for supporting the service. As a side result, it has been shown that if a constant probability of demonstrating quality degradation is included in the providers' payoff functions for the payment-partition game, this does not affect the optimal partition proposed by the Nash bargaining solution, even though it affects the payoff of the individual networks. This is important since the technology employed by each provider and the established data protection mechanisms are always susceptible to malicious attacks due to the dynamic nature of the network and the mobility of the users.

Once the optimal partition is determined, the payment-partition game is modelled as a one-shot *Bayesian* game, to investigate truthfulness on behalf of the participating providers regarding their own costs, since the declaration of costs is very important in determining the optimal partition. It has been shown that no matter whether a provider believes that the interacting provider has declared lower or higher cost than its own cost, it is still motivated to lie about its real costs. To motivate truthfulness, a *pricing mechanism* can be used, which works very effectively towards motivating the providers to be truthful.

References

1. Antoniou, J., Koukoutsidis, I., Pitsillides, A., Stavrakakis, I.: Access network synthesis in next generation networks. Comput. Netw. **53**(15), 2716–2726 (2009)
2. Aumann, R.J.: Acceptable points in games of perfect information. Pac. J. Math. **10**, 381–417 (1960)
3. Aumann, R.J.: Game Theory in the Talmud. Jewish Law and Economics Research Bulletin Series (2003)
4. Axelrod, R.M.: The Evolution of Cooperation. Basic Books, New York (1984)
5. Cesana, M., Malanchini, I., Capone, A.: Modelling network selection and resource allocation in wireless access networks with non-cooperative games. In: Proceedings of the 5th IEEE Mobile Ad-Hoc and Sensor Systems (2008)
6. Cesana, M., Gatti, N., Malanchini, I.: Game theoretic analysis of wireless access network selection: models, inefficiency bounds, and algorithms. In: Proceedings of the 2nd International Workshop on Game Theory in Communication Networks (2008)
7. Crawford, V.P.: Equilibrium without independence. J. Econ. Theory **50**(1), 127–154 (1990)
8. Darwin, C.: The Descent of Man and Selection in Relation to Sex, vol. 1. John Murray, London (1871)
9. Dixit, A., Skeath, S.: Games of Strategy. W.W. Norton & Company, New York (1999)
10. Dyer, J.S., Fishburn, P.C., Estever, R., Wallenius, J., Zionts, S.: Multiple-criteria decision-making: multiattribute utility theory - the next ten years. Manage. Sci. **38**(5), 645–653 (1992)
11. Elayoubi, S.E., Chahed, T., Hebuterne, G.: Admission control in UMTS in the presence of shared channels. Comput. Commun. **27**(11), 1115–1126 (2004)
12. Fitzek, F.H.P., Katz, F.H.P.: Cooperation in Wireless Networks: Principles and Applications. Springer, Berlin (2006)
13. Gao, D., Cai, J., Ngan, K.N.: Admission control in IEEE 802.11e wireless LANs. IEEE Netw. **19**(4), 6–13 (2005)
14. Gintis, H.: Game Theory Evolving: A Problem-Centered Introduction to Modeling Strategic Interaction. Princeton University Press, Princeton (2000)
15. Harsanyi, J.C.: Games with incomplete information played by Bayesian players. Behav. Sci. **14**, 159–182 (1967)
16. Iera, A., Molinaro, A., Campolo, C., Amadeo, M.: An access network selection algorithm dynamically adapted to user needs and preferences. In: Proceedings of the IEEE International Symposium on Personal, Indoor and Mobile Radio Communications (PIMRC'06) (2006)
17. Kar, K., Sarkar, S., Tassiulas, L.: A simple rate control algorithm for maximizing total user utility. In: Proceedings of the Twentieth Annual Joint Conference of the IEEE and Communications Societies, INFOCOM (2001)
18. Kendall, G., Yao, X.: The Iterated Prisoner's Dilemma: 20 Years On. Advances in Natural Computation Book Series, vol. 4. World Scientific Publishing, Singapore (2009)
19. Muthoo, A.: Bargaining Theory with Applications. Cambridge University Press, Cambridge (2002)
20. Nash J.F.: The bargaining problem. Econometrica **18**(2), 155–162 (1950)
21. Nisan, N., Ronen, A.: Algorithmic mechanism design. Games Econom. Behav. **35**(1–2), 166–196 (2001)
22. Niyato, D., Hossain, E.: A cooperative game framework for bandwidth allocation in 4G heterogeneous wireless networks. In: Proceedings of the IEEE International Conference on Communications 2006 (ICC '06), pp. 4375–4362 (2006)
23. Noam, E.: Interconnecting the Network of Networks. The MIT Press, Cambridge (2001)
24. Rubinstein, A.: Perfect equilibrium in a bargaining model. Econometrica **98**(1), 97–109 (1982)
25. Schelling, T.C.: The Strategy of Conflict. Harvard University Press, Cambridge (1960)
26. Selten, R.: Reexamination of the perfectness concept for equilibrium points in extensive games. Int. J. Game Theory **4**, 25–55 (1975)
27. von Neumann, J., Morgenstern, O.: Theory of Games and Economic Behavior. Princeton University Press, Princeton (1944)

Chapter 2
Using Game Theory to Address New Security Risks in the IoT

Abstract An Internet of things (IoT) botnet is a network formed by connecting together a number of computers, smart devices and appliances connected to the Internet. This new type of network does not come without risk, since devices participating in the IoT can easily be infected with malicious software in order to be controlled by authorities that are not the owners of the devices. The botnet is in turn used, without the knowledge of the devices' owner(s), and forced to act as a set of transmitting devices on behalf of the hackers. It is often the case that owners of smart devices and IoT-connected appliances do not consider the significance of having strong security when connecting these devices to the Internet, even if it is as simple as making sure that a strong password is set. It is very important to note here that strong passwords that are updated often can quite satisfactorily safeguard access to the device. The chapter uses game theory to model a coalition that would handle the password generation process in such an IoT scenario, in order to reduce the risk of IoT nodes being hijacked.

Keywords Internet of things · Botnets · Password generation service · Game theory · Coalition formation

2.1 Introduction of a Security Scenario in IoT Communications

An Internet of things (IoT) botnet is a network formed by connecting together a number of computers, smart devices and appliances connected to the Internet. The *things* connected to the Internet often have no user interface, and exchange data based on automated processes without any user intervention. This new type of network does not come without risk. In fact, one of the main risks is that the

© Springer Nature Switzerland AG 2020
J. Antoniou, *Game Theory, the Internet of Things and 5G Networks*,
EAI/Springer Innovations in Communication and Computing,
https://doi.org/10.1007/978-3-030-16844-5_2

devices can be hacked, in order to include them in a *botnet*, i.e. a network of devices infected with malicious software in order to be controlled by authorities that are not the owners of the devices. The botnet is in turn used, without the knowledge of the devices' owner(s), and forced to act as a set of transmitting devices on behalf of the hackers. Once hijacked the devices participating in a botnet are referred to as *bots*.

Nevertheless, note that the use of the word bot to refer to an interconnected smart device in the IoT does not always imply malicious behaviour. For example, Microsoft Azure [7] advertises an IoT bot that makes it easy for the user to control devices around the house, using voice or interactive chat commands. The company claims that *people love to talk to their things* and to *affect their environment*.

Turning our attention back to the malicious scenario, we will aim to have a more detailed look into the attack in order to try and model possible strategic behaviour that can be adopted by the IoT nodes, using game theory. The attack that has the purpose of hijacking IoT devices, and turn them into bots, can be elaborated in the following way:

> The attacker(s), activates malicious software in the Internet, which scans Internet-connected Local Area Networks and IoT local networks for connected devices. Once the devices are detected, the software scans them for active IP addresses. Any device or appliance connected to the IoT would have an active IP. Once an active IP is found then by using available software for finding usernames and passwords, the device is broken into and this is reported back to the attacker, specifically to the software that controls the process. This software will then forward to the hacked device a copy of the malware for hijacking the device into joining the botnet. This continues one device at a time, until the botnet is formed. The botnet can then be controlled to send traffic to specific targets. Even though each infected device sends a small amount of traffic, often due to its own limited capabilities, the botnet itself generates sufficient traffic to potentially carry out, potentially harmful, Denial of Service attacks.

Although such an attack can be carried out on any device connected to the traditional Internet or the IoT, those devices whose owners are not security-aware are more susceptible to be attacked. It is often the case that owners of smart devices and IoT-connected appliances do not consider the significance of having strong security when connecting these devices to the Internet, even if it is as simple as making sure that a strong password is set. It is very important to note here that strong passwords that are updated often can quite satisfactorily safeguard access to the device. A denial of service attack can thus originate from a botnet of many such bots and result in the system being attacked to perform badly, e.g. to slow down, or even to suffer a system crash.

Now, let's assume that the attacker trying to recruit a specific device in an IoT botnet will randomly look for an active IP and once the active IP is found that the controlling software will try to hijack other *nearby* devices to strengthen the transmission capabilities [11], either as part of the initial attack or by using the newly infected devices to hijack their own neighbours. We may assume that once a device has been hijacked into a bot, it can act as a malicious node and attack its neighbours, since IoT devices often don't have the capabilities to have the malware removed.

Given the above-mentioned information, the chapter proposes that, in order to prevent such malware attacks, the IoT devices must automatically be configured to frequently change their passwords, and always use strong, not recently used passwords. It is safe to assume here that such connected IoT devices are often characterized by limited computational power, and that no single device on its own can handle the anticipated demand of applying the selected security measures for a certain period (e.g. days, weeks, months, etc.). Therefore, to implement the required security measures, the chapter proposes that nearby IoT nodes can form a coalition, such that they use their joined resources to safeguard against such attacks.

Potentially, many different combinations of nearby nodes can jointly provide sufficient resources to meet demands of employing a group password mechanism, responsible to regenerate passwords for the IoT nodes that are strong and not recently used. It may be assumed that for each IoT neighbourhood or group of nearby nodes, a subset of these nodes is sufficient to support the required security measures. The following paragraphs and sections in this chapter show how game theory can be used to choose such a subset of nodes. It is significant to choose the subset that may provide the minimum required amount of resources, by engaging the least number of nodes possible in order to avoid relying on a large number of nodes for each password change. The reason that a large number of nodes are not desired in this situation is that employing a large number of nodes into the process of password generation can result in unwanted delays due to the limited computational resources of each one node on its own, as there will need to be a distributed coordination effort (Fig. 2.1).

2.1.1 Scenario on Password Generation for IoT Nodes

Consider the scenario where the IoT nodes cooperate to satisfy the need for accommodating a multiparty security service for the IoT group of *collocated* or *nearby* nodes. Given that all participating IoT nodes will benefit from the employed password generation mechanism, a selection of the *stronger* nodes in the group must be made, such that the employed mechanism can sufficiently serve the group of IoT nodes, i.e. engage sufficient resources but avoid detrimental communication delays.

The particular scenario considers that it is the case that none of the participating IoT nodes can support the group in its entirety, by itself. Node cooperation can ensure a more efficient resource planning and can support the security service for all users, by having the IoT nodes form coalitions and thus collectively managing to provide the total resources necessary to support the password service. Note that when we refer to the security or the password service in this chapter, we are referring to the same mechanism of password generation for the IoT nodes. An attacker will be unaware of this cooperation, even though the attacker will be aware that there exist frequently generated passwords across the group of nodes.

Fig. 2.1 The IoT nodes are interconnected, through their connections to the Internet, and can be found in objects such as cars or electronic devices, but also on user devices, e.g. smart mobile devices

Fig. 2.2 A subset of the nearby IoT nodes can form a coalition to join resources in order to support the password generation mechanism that may provide security against malware attacks for the collective group of IoT nodes

The nodes themselves will be indifferent to the subset of nodes responsible to carry out the password service, as long as the mechanism is successfully employed across the group. The incentive of the nodes that end up participating in the password generation coalition is discussed next (Fig. 2.2).

2.1.2 Incentive for the IoT Nodes to Participate

It is important to note that the IoT nodes themselves have an incentive to participate in the password generation coalition, in order to eliminate the risk of being hijacked, which is much greater for a single node than for a coalition of nodes. Since the participating nodes will be generating the initial passwords (including the necessary password updates), participating in the subset of nodes involved in this process implies that the participating nodes themselves will be the first to receive their updated passwords. Minimizing the quantified risk of being hijacked is considered to be the basis of the payoff allocation in the proposed scenario. The payoff allocation to the members of the formed coalition is a consideration for all the IoT nodes participating in the coalition formation process. Given that each IoT node participating in a coalition is solely motivated by its need to minimize its risk of being hijacked, we further investigate how such coalitions are formed, so that they satisfy the whole IoT group at the same time. For this purpose we make use of the theoretical tools of coalitional game theory, and subsequently, coalitional games.

Coalitional games deal with the situation, in which interactions occur between groups of players and not individual players. The term *coalition* refers to a group of game players. Therefore, actions are assigned to coalitions even though individual entities may consider their own preferences, especially when selecting a particular coalition in which to participate. A coalitional model is characterized by its focus on what groups of players can achieve rather than on what individual players can achieve.

In coalitional games, the way a coalition operates internally, i.e. among its members, is not part of the game definition as it is considered an important aspect for a coalitional game. The reason is that the outcome does not depend on such details. A solution concept for coalitional games assigns to each game a set of outcomes, capturing consequences for the participating coalitions. The solution defines a set of arrangements that are stable in some sense, i.e. that the outcomes are immune to deviations of any sort by groups of players.

In order to determine the solution to a coalitional game, the way payoffs are assigned to the various coalitions must be defined. Such payoff assignment can occur per group as a whole, or per group using a particular division arrangement within the group for its members. When payoffs are assigned per group, the players that participate in the same group are associated with the group's payoff and it is not defined how this payoff may be further partitioned among its members. This case of payoff assignment is referred to as transferrable payoff coalitional game. The alternative is known as non-transferable payoff coalitional game, and in such model there exists a rule on how group payoffs are divided among participating players. The subsequent analysis will provide mathematical definitions for each type of payoff.

A final note on introducing the concept of coalitional game models is that there exists a well-known solution concept for a coalitional game model, often used to quantify the value of each participating coalition. This solution concept is known as the *core*. The core is a solution concept that requires that no set of players be able to break away and take a joined action that makes all of them better off. Overall, the idea of the *core* is analogous to that behind a Nash equilibrium of a non-cooperative game, i.e. an outcome is stable if no deviation is profitable. In the case of the *core*, an outcome is stable if no coalition can deviate and obtain an outcome better off for all its members.

2.2   Coalition Selection Game

Consider that the nearby IoT nodes are under different administration authority or ownership, and therefore, the decision of whether to participate in a certain resource combination or not would be shaped by each node's goal to minimize its risk of being hijacked. As a result, a coalition game would arise in this environment, with each IoT node in the group aiming to participate in a *prevailing*, otherwise referred to as a *winning* coalition (the one that will be selected). Participating in the winning coalition would yield the largest possible benefit to an IoT node, as this would result in its own resources being protected from a potential future attack.

The formation of coalitions depends on the available resources of each IoT node, such that the collective amount of resources is sufficient to support the password generation mechanism required. In addition to the required resources, the formation of the coalitions depends on the quantifiable payoffs that will be allocated to the participating nodes, such that there exists an incentive for the nodes to commit their available resources towards this group mechanism.

The section introduces and defines a coalitional game, where out of several coalitions that can be formed, the *best* one is selected, according to our defined criteria. We henceforth refer to this game as the *coalition selection game*. The nodes participating in the coalition selection game aim to maximize their payoff by participating in the selected coalition. We consider for this game, a payoff allocation approach according to values of (normalized) power indices, i.e. numerical values that are used to measure the influence of a node on the formation of coalitions and thus on the game itself. Payoffs are thus determined based on the power of each node in the game, i.e. its index. The reason for this is that since the payoff relates to the minimization of the risk of being hijacked and this depends on the selected coalition, then by having a greater contribution in the selected coalition, a node achieves more control over the security mechanism and thus experiences less risk.

To allocate indices to the participating nodes, we consider well-known indices, such as the Shapley–Shubik index [10] Banzhaf index [2], the Holler–Packel index [6], and the popularity power index [1], which is more appropriate for the particular coalition selection scenario because it associates the popularity of each IoT node to the number of stable coalitions it participates in, and thus can better represent the group of IoT nodes by participating in the selected coalition.

Following in the chapter, definitions of such terms as *stable*, when we refer to a formed coalition, will be given. The popularity power index aims to achieve fairness, in the sense that it only considers the possible coalitions that would be formed if payoffs were assigned proportionally to the players' (i.e. the IoT nodes) contributions, i.e. in a fair manner. Given the payoffs relate to the risks of nearby IoT nodes being hijacked, then, as previously mentioned, the more the control over the coalition (i.e. the more resource contribution by a single node), then the less the risk experienced by the particular node.

The evaluation presented in the subsequent sections of the chapter focuses on the use of the aforementioned indices, with particular attention to the PPI. The chapter later evaluates the PPI by introducing an analysis of the stability of coalitions according to the core and inner core concepts [5], considering both transferable and non-transferable payoffs, i.e. payoffs that do not vary according to resource contribution to the coalition and payoffs that do. Furthermore, we show that the coalitions that are formed when considering the PPI are only coalitions that are stable.

2.2.1 Definition of the Coalition Selection Game

Next we define the game model, to be used as guide for the payoff allocation strategies.

Definition 2.1 (The Coalition Selection Game) Let $\mathcal{N} = 1, 2, \ldots, N$ denote the set of game players, i.e. the IoT nodes, and let S denote the set of all possible coalitions, i.e. the set of all non-empty subsets of $\mathcal{N}$. Let R denote the least amount of resources needed for accommodating password generation demands, and let r_i denote the amount of available resources of the ith member of the coalition (members can be ordered arbitrarily). We may assume that the available resources of each member are known to all members of the coalition. Note that although in such environments, the calculation of available resources may be a difficult task, e.g. in a wireless environment signal interference may be present, or, a node may wish to distort resource information made public. However, for the purposes of this model we assume that appropriate policies are in place, even though a detailed study of these policies will not be explored here.

The characteristic function of the game is

$$v(S) = \begin{cases} 1, & \text{if } \sum_{i=1}^{|S|} r_i \geq R \\ 0, & \text{otherwise .} \end{cases} \tag{2.1}$$

That is, a coalition has positive value only if the sum of available resources of its members is greater or equal to the resource threshold R. This definition corresponds to a simple (or 0–1) game; the game is also monotonic since $v(S_1) \leq v(S_2)$ for all $S_1 \subseteq S_2$. A coalition S is said to be *winning* if $v(S) = 1$, otherwise it is said to be *losing*, i.e. the collective resources of its members are not equal or greater than the total amount of required resources. A player $i \in S$ is said to be a *null player for coalition S* if $v(S) = v(S \setminus \{i\})$. It is generally called a *null player* if this holds for every coalition S to which it may belong. To avoid trivialities, we will generally assume that $r_i < R \ \forall i \in \mathcal{N}$, and that $\sum_{i=1}^{N} r_i \geq R$.

Candidate solutions to the game are referred to *minimal winning coalitions*. A winning coalition is said to be minimal if it becomes a losing one upon departure of any member. A related notion is that of a *by-least winning coalition*. In [4], the authors define a coalition S to be *least winning* if it is minimal winning and for any other minimal winning coalition S' it holds that $W(S) \leq W(S')$. A related definition is introduced here from the point of view of each player $i \in S$. Therefore, we denote by $W(S) = \sum_{i=1}^{|S|} r_i$ the sum of the available resources of the members of S, a coalition S is said to be *by-least winning with* (or *for*) player i, if it is a minimal winning coalition, it contains i, and for any other minimal winning coalition S' containing i it holds that $W(S) \leq W(S')$; thus, the by-least winning coalition is the best for the node i to participate in, out of all possible winning coalitions it can participate in.

2.2.2   Definitions of the Power Indices

The formation of a coalition is greatly influenced by the players themselves in terms of how much they motivate such cooperation with other players. In the password generation scenario, it is assumed that such motivations for cooperation exist within the group of nearby IoT nodes, because of the need for a collective solution that prevents potential hijacking of the IoT nodes (by the use of malware propagated through the group). The solution is based on the intelligent use of passwords by employing a password generation mechanism. Although the details of the algorithm itself are out of the scope of this chapter, it is important to understand the need for collective resources to support the implementation of this mechanism, as it is not expected that any one IoT on its own can support its implementation.

For simple (0–1) games, a popular coalition formation method is by using power indices, briefly mentioned in the previous section. A power index (or value) is commonly used to measure the influence of a player on the formation of coalitions and most importantly on the outcome of the game. The notion of power indices can be used as a solution concept for the game itself, since, if no payoff allocation rule is specified a priori, then normalized power indices can be used to allocate payoffs. Alternatively, this approach can be used in cases where the payoff allocation is determined by the use of a common pool in which IoT nodes share their resources, i.e. in the case they are controlled by the same authority, and there is no coalition formation process by the actual IoT nodes themselves. Nevertheless, as we may not assume that the nodes are all under the same authority, we approach the proposed solution as a coalition formation approach.

Widely used power indices are the *Shapley–Shubik power index* (abbreviated here as SSPI) [10], and the *Banzhaf power index* (abbreviated here as BPI) [2]. These are generally sums of the *marginal contributions* $(v(S) - v(S \setminus \{i\}))$ of a player i to each coalition, weighted by different probability distributions over the set of coalitions. Thus, a player i is *critical* to coalition S if its marginal contribution is 1; otherwise, it is non-critical.

The SSPI assumes all permutations of the order that members form a coalition are equally likely [1], and is defined by:

Definition 2.2

$$\text{SSPI}_i(N, v) = \sum_{\substack{S \subseteq \mathcal{N} \\ (S \ni i)}} \frac{(|S| - 1)!(N - |S|)!}{N!} (v(S) - v(S \setminus \{i\})) . \qquad (2.2)$$

The BPI assumes, on the other hand, that all possible coalitions that contain i are equally likely [1], and is defined by:

Definition 2.3

$$\text{BPI}_i(N, v) = \frac{1}{2^{N-1}} \sum_{\substack{S \subseteq \mathcal{N} \\ (S \ni i)}} (v(S) - v(S \setminus \{i\})) , \qquad (2.3)$$

for $i = 1, \ldots, N$.

Another popular power index based on minimal winning coalitions is the *Holler–Packel power index* (HPI) [6]. For simple games, the HPI is defined as:

Definition 2.4

$$\text{HPI}_i(N, v) = \sum_{S \in M(N,v)} (v(S) - v(S \setminus \{i\}))$$
$$= |\{S \in M(N, v) : i \in S\}| , \qquad (2.4)$$

for $i = 1, \ldots, N$, where $M(N, v)$ is the set of all minimal winning coalitions.

We define also the normalized *Holler–Packel value* $\overline{\text{HPI}}_i = \text{HPI}_i / \sum_{i=1}^{N} \text{HPI}_i$, which represents the that proportion of minimal winning coalitions player i is in [1].

An important property of both the SSPI and the BPI indices in monotonic simple games is that players with greater weight (contribution) also get a greater index. This is evident here since if a player i is critical to a coalition $S \cup \{i\}$, then a player i' with $b_{i'} > b_i$ is also critical to coalition $S \cup \{i'\}$.

2.2.3   The Equivalence of the Coalition Selection Game for the Password Generation Scenario with Classic Game Theoretic Models

This section demonstrates the equivalence of the coalition selection game for the password generation scenario to the weighted voting game, a well-studied paradigm and game theoretic model from which many useful conclusions can directly apply upon the studied scenario.

Definition 2.5 A weighted voting game consists of N players and a weight vector $w = (w_1, w_2, \ldots, w_N)$, where w_i reflects the *voting weight* of player i. Let $W = \sum_{i=1}^{N} w_i$. For a coalition S, the characteristic function of the game is

$$v(S) = \begin{cases} 1, & \text{if } \sum_{i=1}^{|S|} w_i > \frac{W}{2} \\ 0, & \text{otherwise .} \end{cases} \tag{2.5}$$

We assume $w_i \leq W/2 \ \forall i \in \mathcal{N}$.

Proposition 2.1 *The weighted voting game can be mapped onto a coalition selection game as this may be applied to a password generation scenario with networks' resources $r_i = w_i$ and minimum required resources to accommodate a service equal to $R = \frac{W}{2}$.*

Proof To prove the equivalence, it suffices to show that there exists a one-to-one mapping between vectors w and $r = (r_1, \ldots, r_N)$ so that $\sum_{i \in S} w_i > W/2$ if and only if $\sum_{i \in S} r_i \geq R \ \forall S \subseteq \mathcal{N}$. A mapping satisfying these requirements is readily obtained by setting $w_i = r_i/2$ and W to a number R^* arbitrarily close to R such that $\sum_{i \in S} r_i \geq R$ iff $\sum_{i \in S} r_i > R^*$. (Mathematically, it is straightforward to postulate that such a number does exist, since the values represented by r_i are discrete.)   □

In addition to the above observations, based on the equivalence of the coalition selection game to the weighted voting game, two additional observations are made in [3, 4] with regard to the behaviour of the power indices defined in Sect. 2.2.2: firstly, that restricting our attention to minimal winning coalitions as with the HPI results in weaker players would result in IoT nodes with a relatively smaller amount of available resources, getting higher power, compared to the measurement with the SSPI and the BPI, and, secondly, that with the HPI the monotonicity of the players' power indices to their weights may not be preserved, i.e. an IoT node with smaller weight may get a higher HPI ranking than an IoT node with greater weight (note that the weights are directly related to the amount of available resources of each node).

2.2.4 Payoff Allocation

In the coalition selection game, we seek for the most possible winning coalition to support a particular resource demand, i.e. the coalition that is most likely to be formed, so that the allocation of payoffs is more fair. Nearby IoT nodes participate in a coalition and offer their resources in return for more security or less risk; thus, the payoff represents the amount of security achieved or the minimization of risk of a node to be hijacked. For example, the reward for allowing their available resources to be reserved, in order to be used to support the password generation mechanism, can include more sporadic security scans or in cases where IoT nodes don't undergo security scans, the reward is the capability to guarantee a less risky implementation to their owners. It is considered that all IoT nodes are independent and behave rationally, and that the objective of each IoT node is to maximize its payoff, i.e. to minimize its risk of being hijacked.

Clearly, which coalition(s) will finally be formed depends on how payoffs are allocated. Moreover, if coalitions are formed arbitrarily and each IoT node could participate in more than one coalition, then the criterion for selecting which coalition to participate in is the payoff received from each, naturally preferring the highest. This preference should be captured in the power index used to allocate payoffs, i.e. the power index should be defined in a way that the most *popular* coalition, i.e. the one preferred by most players, is favoured. Initially, however, we will concentrate on payoffs and consider payoff allocations for the game without the use of power indices, and, subsequently, we will return to a discussion of power indices.

Assuming that no power indices are used, and that payoffs are assigned based on coalition formation, let the total payoff allocated to the set of players be T and the payoff allocation vector be $t = (t_1, t_2, \ldots, t_N)$, such that $t_i \geq 0 \ \forall i = 1, \ldots, N$ and $\sum_{i=1}^{N} t_i = T$; an allocation satisfying the above conditions is said to be *feasible*.

We may consider two types of payoff, according to coalitional game theory: (a) *transferable* payoffs between the participating IoT nodes, and (b) *non-transferable* payoffs. In the transferable payoff case, individual IoT nodes can transfer any portion of their payoff to other members of the coalition, as long as their final payoff remains greater than zero. These transfers may be viewed as *side payments*, in this particular case, serving the needs for password generation of a node ahead of one's own similar needs. Such *side payments* may be used as a means to *attract* other players, i.e. IoT nodes, in a specific coalition. In the non-transferable case, such side payments are not allowed and we will consider that IoT nodes attain a payoff that is proportional to their resource contribution, relative to the other members of the coalition, by always serving their own needs for password generation first. More specifically, if this winning coalition is $\mathcal{K}$ consisting of $K \leq N$ IoT nodes with available resources $r_1, r_2, \ldots, r_K$, then

$$t_i = \begin{cases} \frac{r_i}{\sum_{j=1}^{K} r_j} T, & \text{if } i \in \mathcal{K} \\ 0, & \text{otherwise .} \end{cases} \tag{2.6}$$

(It holds that $\sum_{j=1}^{K} r_j \geq R$.)

When payoffs are transferable, this leads to trivial solutions of minimal-sized coalitions. When payoffs are non-transferable, the proportional payoff allocation case is more interesting and requires the notion of a *by-least winning* coalition. Given the defined power indices, we may relate HPI to minimal winning coalitions; however, none of the defined power indices relates directly to the concept of by-least winning coalition. In [1], the authors, having looked at the possible solution concepts and recognizing the lack of alignment of a particular power index with the notion of a *by-least winning* coalition, proposed PPI, which we adopt here as an appropriate solution concept for the coalition selection game of the password generation scenario. We thus explain the corresponding stability analysis for the selected power index to show how the PPI relates to coalitions that are stable.

2.3   A Stability Analysis for the Popularity Power Index

The stability analysis aims to determine a *fair* way of allocating payoffs to a subset of the IoT nodes, so that a stable coalition is formed and the desirable password generation service is provided. Although it could be argued that every minimal winning coalition could potentially be a solution of the game, the stability analysis shows that the set of possible solutions can be further reduced, since not all minimal winning coalitions are equally likely. Rather, each IoT node participating in the game has specific preferences, i.e. to end up with higher payoff, and to be in one or more coalitions, which are by-least winning with it.

Let $M(N, v)$ denote the set of all minimal winning coalitions and let $Z_i(N, v)$ be the set of coalitions which are minimal in size for player i. A coalition S is said to be *minimal in size for* $i \in S$, iff $|S| \leq |S'| \; \forall S' \ni i$, where $S, S' \in M(N, v)$. Considering the proportional allocation rule, if a player i belongs to two minimal winning coalitions S and S', then $(r_i / \sum_{j \in S} r_j)T > (r_i / \sum_{j \in S'} r_j)T$ if $\sum_{j \in S} r_j < \sum_{j \in S'} r_j$, and hence it would get a higher payoff in the by-least winning coalition. We denote by $L_i(N, v)$ the set of all by-least winning coalitions for player i.

The popularity power index is based on the popularity of all coalitions which are in $\cup_{i=1}^{N} L_i(N, v)$ (a subset of $M(N, v)$). For each minimal winning coalition $S \in M(N, v)$, we define as its *preference index* $\omega(S)$ the total number of preferences it gathers by all players:

$$\omega(S) = |\{i \in \mathcal{N} : S \in L_i(N, v)\}| . \tag{2.7}$$

Thus, the PPI is defined as follows:

$$\text{PPI}_i(N, v) = \sum_{S \in M(N,v)} \frac{\omega(S)}{\sum_{k \in M(N,v)} \omega(k)} I_{iS} , \qquad (2.8)$$

where I_{iS} equals 1 if $i \in S$ and 0 otherwise. Simply put, the index PPI_i equals to the probability that if a coalition would be selected by asking one player in $\mathcal{N}$ randomly (and further, if when this player had multiple equal preferences, he would select one of them with equal probability), then a winning coalition would be selected that contains player i. Hence, this index relates the popularity of minimal winning coalitions a player belongs in, to this player's power. As with the other indices, define in [1], a normalized form of PPI is also defined as follows: $\overline{\text{PPI}}_i = \text{PPI}_i / \sum_{i=1}^{N} \text{PPI}_i$.

In Sect. 2.3.1, we relate PPI to the stable coalitions formed in case no power indices are used, and payoffs may be either transferrable or non-transferable and proportionally allocated.

2.3.1   The Concept of the Core for the Coalition Selection Game

In this section, the well-known concept of the *core* (see, e.g. [8]) is used to examine the stability of coalitions formed according to the payoff allocations considered if no power index is used to allocate payoffs. The concept will be discussed and subsequently, the PPI will be shown to be equivalent to this concept with regard to payoff allocation for stable coalitions, since we aim to show a solution for the coalition selection game.

Specifically, we may define a payoff allocation to a set of N players to be in the core of a coalitional game if there is no other coalition wherein each member can get a strictly higher payoff than the payoff received by participating in the specific allocation. Such an allocation, as well as the coalitions that it induces, can be referred to as stable, since there would not be any motivation to break the specific coalition in order to form other ones by the participating players.

In order to apply the core concept, we redefine the characteristic function of the coalition selection game, found in Eq. (2.1) in the transferable payoff case to the following, to refer to the payoff T:

$$v(S) = \begin{cases} T, & \text{if } \sum_{i=1}^{|S|} r_i \geq R \\ 0, & \text{otherwise} . \end{cases} \qquad (2.9)$$

The redefined function states that when the minimum necessary resources are available, the value of the characteristic function equals the total payoff.

For the transferable payoff case we are not interested in how the payoff is divided among the members of the coalition, and therefore, an allocation is in the core of the transferable payoff game—specifically, the coalition selection game—if it does not give an incentive to the participating IoT nodes to deviate, i.e. to break the coalition, and obtain an outcome better for all the members of the selected coalition.

Definition 2.6 An allocation $t = (t_1, t_2, \ldots, t_N)$ is said to be in the core of the coalition selection game of the password generation scenario, when the case of transferable payoffs is considered iff (if and only if),

$$\sum_{i \in \mathcal{N}} t_i = T \text{ and } \sum_{i \in S} t_i \geq v(S), \ \forall S \subseteq \mathcal{N} \ .$$

Since $v(S)$ takes either the value 0 or T in the coalition selection game according to the definition of the payoff function, this reduces to the requirement that for every winning coalition that could be formed by the networks, the sum of payoff allocations should always be equal to T.

For the non-transferable payoff case, we have the following definition:

Definition 2.7 An allocation $t = (t_1, t_2, \ldots, t_N)$ is said to be in the core of the coalition selection game of the password generation scenario, when the case of non-transferable payoffs is considered iff (if and only if), $\sum_{i \in \mathcal{N}} t_i = T$ and there exists no other payoff allocation $y = (y_1, y_2, \ldots, y_N)$ derived according to (2.6) for which $y_i > t_i, \ \forall i \in S \subseteq \mathcal{N}$, for any $S \subseteq \mathcal{N}$.

That is, the allocation must satisfy the requirement that the total amount of resources is greater or equal to the required resources for supporting the password generation mechanism, and in addition, that there exists no other allocation which gives strictly higher payoff to any of the participating IoT nodes that are members of the specific coalition.

It follows that a single winning coalition is mapped to an allocation in the core of the game. In the non-transferable payoff case, this is defined to be the coalition $\mathcal{K}$, based on which the payoff vector is derived. In the transferable payoff case, this is defined as $\{i \in \mathcal{N} : t_i > 0\}$, the set of players with positive payoff. We can then refer to *coalitions in the core*, as the set of winning coalitions for which their corresponding allocations are in the core.

Moreover, we deduce that not all winning coalitions are in the core. In fact, we have the following:

Theorem 2.1 *In both the transferable and non-transferable payoff cases defined above, only minimal winning coalitions are in the core.*

Proof In the transferable payoff case, notice that for any non-minimal winning coalition, a corresponding minimal one can be formed by the players that are non-null (in the coalition). Then for any payoff allocation to players in the non-minimal winning coalition, the players in the minimal winning coalition can divide the excess payoff in such a way that they all get strictly higher payoff. In the non-transferable payoff case, the statement of the theorem follows directly from (2.6).

Remark   This theorem follows from Riker's *size principle* [9], which was shown for weighted voting games in [3] and can hence be applied to the coalition selection game.

Given the above, for the coalition selection game, we will be focusing primarily on minimal winning coalitions, $M(N, v)$. Note that although the characteristic function v is defined usually for transferable payoff games, it can also be used in this discussion in the context of the non-transferable payoff game such as $M(N, v)$ and $L_i(N, v)$, since, along with (2.6), it can be used to define the specific coalition selection game.

Looking back into the transferable payoffs case, it follows that only minimal winning coalitions which are also minimal in size for at least one of their members, denoted as $Z_i(N, v)$, will be in the core of the coalition selection game. In fact, when considering transferable payoffs, all minimal winning coalitions of the same size can be treated as equivalent, since the payoff is considered for the selected coalition in its totality and not considered for individual members.

However, considering the non-transferable payoff case, then only coalitions that are by-least winning for at least one player, denoted by $L_i(N, v)$, or simply by L_i, are solutions in the core of the game. Therefore, we must proceed to investigate which coalitions are in the *inner core* of the game. The inner core [8] is a subset of the core that contains coalitions that are even more stable, in the sense that there exists no *randomized plan* that could prevent their formation.

Definition 2.8   A randomized plan is any pair $(\eta(S), y(S))$, $S \subseteq \mathcal{N}$, where η is a probability distribution on the set of coalitions, and y is the vector of payoff allocations for the members of coalition S, $y(S) = (y_i(S))_{i \in S}$. For non-transferable payoff games, the inner core is composed of all allocations t (or corresponding coalitions) for which $\sum_{S \supseteq \{i\}} \eta(S) y_i(S) < \sum_{S \supseteq \{i\}} \eta(S) t_i$, for some $i \in \mathcal{N}$, in all randomized plans $(\eta(S), y(S))$.

To better understand the inner core concept, and how it can be used in the coalition selection game, consider the following example. In the password generation scenario, the coalition selection game may represent a centralized controller for the nearby IoT nodes that selects the coalition, which will be S_1 with probability t_1, or S_2 with probability $t_2 = 1 - t_1$, so that it is stable. In order for the coalition to be stable, the payoff given to each network should not be smaller than the mean payoff anticipated in the coalition created by the centralized controller. Since nearby IoT nodes may not have the option of such a centralized controller, then they are expected to arrive to this payoff allocation agreement on their own. This is further demonstrated in the following theorem.

Theorem 2.2 *In the coalition selection game of the password generation scenario, for the case of non-transferable payoffs, where these payoffs are proportional to the available resources of the participating IoT nodes, all coalitions which are by-least winning for at least one of their members are in the inner core of the game.*

Proof In the coalition selection game, if a participating IoT node also participates in several coalitions that are by-least winning with it, then in the non-transferable payoff case it would get the highest possible reward in every one of these coalitions. This reward would further be the same in every randomized plan among these coalitions, and lower for randomized plans containing coalitions other than the by-least winning.

Since, for a player i, the allocation in a by-least winning coalition is the maximum it can get, there exists no randomized plan that could block these coalitions from forming and hence the latter are in the inner core of the game. □

Remark Of all by-least winning coalitions which are in the inner core, we may conclude that those that are by-least winning for *all* their members are *most stable*.

The formalization of this conclusion referred to as the concept of *stability under uncertainty of formation* [1] is presented in Sect. 2.3.2.1.

The game can also be further analysed to show that this minimal in size coalition, which is by-least winning, does exist.

Theorem 2.3 *In the coalition selection game with non-transferable payoffs proportional to the available resources of the participating networks, there exists at least one coalition which is by-least winning for all its members. Further, regardless of the payoff allocation, there exists at least one coalition that is minimal in size for all its members.*

Proof The proof presents the above for by-least winning coalitions only since to prove the theorem for minimal in size coalitions is similar because the existence of a by-least winning coalition by definition implies the existence of at least one minimal winning coalition.

Consider the case when there exists a coalition S, such that $\sum_{i \in S} r_i = R$, since then S is by-least winning for all its members. When $\sum_{i \in S} r_i > R$ for some $S \in \cup_{i=1}^{N} L_i$, and there exists $j \in S$ such that $S \notin L_j$, then necessarily another coalition $S_1 \neq S$ exists such that $S_1 \in L_j$ and $\sum_{i \in S_1} r_i < \sum_{i \in S} r_i$. Similarly now, if there exists $k \in S_1, k \neq j$, such that $S_1 \notin L_k$, then there exists another coalition $S_2 \notin \{S_1, S\}$ such that $S_2 \in L_k$ and $\sum_{i \in S_2} r_i < \sum_{i \in S_1} b_i$. Continuing this procedure, since we have a finite number of players, a finite sequence of coalitions $S, S_1, S_2, \ldots, S_m$ is produced, for which $\sum_{i \in S} r_i > \sum_{i \in S_1} r_i > \sum_{i \in S_2} r_i > \cdots > \sum_{i \in S_m} r_i > R$ and S_m is by-least winning for all its members.

It follows that in both transferable and non-transferable payoff cases studied, interests of at least some players coincide. However, it will be reasonable to assume that a player will prefer a minimal winning coalition even though it would not be by-least winning with it, if otherwise it would be excluded from the prevailing coalition and receive zero payoff.

2.3.2 Conflicting Preferences in the Coalition Selection Game

We have so far shown that each IoT node participating in the coalition selection game i would maximize its payoff and hence prefer one of the coalitions in $Z_i(N, v)$ (transferable payoff case) or $L_i(N, v)$ (proportional payoff case) to eventually be formed. Unless $\bigcap_{i=1}^{N} Z_i \neq \emptyset$ in the former, or $\bigcap_{i=1}^{N} L_i \neq \emptyset$ in the latter case, there is no mutually preferred coalition for all participating IoT nodes.

In game theory a model that takes this into consideration is the model of a *coordination game* where at least one player has *conflicting preferences* with one or more of the others, as is expected to be a situation that may arise in the coalition selection game.

A possible resolution of such a coordination game is based on the idea of calculating the probability that these coalitions would randomly form. The analysis applies equally to the transferable and non-transferable payoff cases, and it further uses $\mathscr{G}_i(N, v)$ (or simply $\mathscr{G}_i$) to denote either $Z_i(N, v)$ or $L_i(N, v)$, depending on which of the two payoff cases is under investigation.

2.3.2.1 The Coordination Game Model

The ultimate goal of the game is to find out if one or more stable coalitions can form, so that the password generation service requirements in terms of resources are satisfied. Moreover, what is more important is that, under the proportional payoff allocation rule, if a participating IoT node i can be in multiple minimal winning coalitions, then it prefers the one (or ones) which is (or are) by-least winning with it. As previously shown, in the situation that there are more than one by-least winning coalitions for the same IoT node, then these coalitions must sum to the same total resource amount and hence payoff. For each player i, we denote by $L_i(N, v)$ the set of all by-least winning coalitions for i (this set will also be denoted simply by L_i).

For cases where some players have more than one by-least winning coalitions, we may examine which coalitions are more likely to be formed. For $i = 1, \ldots, N$, we define the *probability of formation* $T_f^{(i)}(S)$ to be the probability that player i would anticipate coalition S to be formed, if player i participated in it and all other players $j \subset S$, $j \neq i$ would independently choose to participate in one of their preferred coalitions with equal probability. That is,

$$T_f^{(i)}(S) = \begin{cases} \prod_{\substack{j \in S, \\ j \neq i}} \frac{1}{|\mathscr{G}_j|}, & \text{if } S \in \bigcap_{j \in S, j \neq i} \mathscr{G}_j \\ 0, & \text{otherwise}. \end{cases} \quad (2.10)$$

Then, given the above, a participating IoT node would ultimately prefer to participate in the coalition which has the highest probability of formation, and hence, under such uncertainty, would potentially offer the highest expected payoff.

The above concept of stability for the winning coalition is referred to as a coalition S that is *stable under uncertainty of formation* if and only if there exists no other coalition S' with a common member with S that anticipates a higher probability of formation for S'. The formal definition is given next:

Definition 2.9 In the coalition selection game with transferable or non-transferable payoffs, a coalition S is stable under uncertainty of formation iff $T_f^{(i)}(S) > 0 \, \forall i \in S$ and,

$$\nexists S' \neq S \text{ s.t } T_f^{(j)}(S') > T_f^{(j)}(S) \text{ for } j \in S \cap S'.$$

Note that the initials s.t. refer to the phrase "such that".

This notion would help to refine solutions of the coordination game model as this is applied to the coalition selection game in case of conflicting preferences of the participating IoT nodes. It also creates a formal ground to confine solutions to coalitions which are minimal in size (transferable payoff case) or by-least winning (proportional payoff case) for all their members: by definition, only such coalitions are stable under uncertainty of formation (in view of (2.10), other coalitions have zero probability of formation for at least one member).

Remark Linking the above concepts to the selected power index for the coalition selection game, namely PPI, we may remark that PPI associates a player's value to the number of stable coalitions it participates in; the higher this number, the greater the index value. In fact the preference denoted by $\omega(S)$, in the definition of PPI, refers to how many players consider coalition S, by-least winning. Clearly, the coalition with the highest ωS is *stable under uncertainty of formation*, according to Definition 2.9.

Thus, given the above definition, it is evident that PPI excludes a number of coalitions which are not stable and hence would not appear if nearby IoT nodes formed coalitions independently. In addition, the PPI is more fair than other indices examined, in the sense that it only considers stable coalitions in the inner core that would be formed if payoffs were allocated proportionally to players' contributions of resources, i.e. in a fair manner. These coalitions have been shown to be the most probable to be formed, both in the transferrable and in the non-transferable payoff cases.

2.4   An Evaluation Example for the Coalition Selection Game

In this section, we propose an example of the coalition selection game by quantifying various indices in order to provide evaluations of different game instances, specifically a game instance with three players and a game instance with four players. The evaluation demonstrates how the PPI allocates payoffs only to the players that would participate in the minimal in size or by-least winning coalition, i.e. players that are in the inner core of the game, as we have shown that all coalitions which are by-least winning for at least one player are in the inner core. The following tables assign numerical behaviour to three and to four players, according to all aforementioned power indices described in the previous sections.

Power index values are examined in the example below, for three players first and subsequently for four players, with different distributions of available resources. Even though the theory extends to many players, we are demonstrating the concept through an instance of the game with three players, and with an instance of the game with four players, only for clarity of illustration and we invite the reader to attempt to recreate the example with a greater number of players.

In terms of resources, the individual resources for each test instance presented in the examples sum up to the same total resource amount. This is necessary in order to better compare results between the different cases, i.e. different instances of player resources for the three and four players examples. Note that collaboration between players is considered only if the sum of their resources adds up to or exceeds the amount of required resources, i.e. one (1) in the example.

In the first example below, the resource distributions D_i ($i = 0, \ldots, 5$) of available resources for each of the three nodes, from the case $i = 0$ where resources are uniformly distributed between access networks, i.e. each IoT node has the same amount of available resources, to non-uniform cases ($i = 1, \ldots, 5$), are carefully selected to exhibit the varying allocations of the power indices, when available resources are about the same, or resources are concentrated in only a few of the IoT nodes. To avoid taking absolute values, we have considered available resources of each IoT node i normalized with respect to the minimum resource requirement R, i.e. r_i/R. For each of the power indices BPI, SSPI, HPI and PPI, we examine both the values of the indices as well as their rankings.

Extending the numerical test cases to include four players instead of three, results in the following:

The values of normalized available resources are shown in Tables 2.1 and 2.3, for the cases of three and four players, respectively. For each one of these cases the

Table 2.1 Example of resource distribution with three players

Distribution	$\frac{r_1}{R}$	$\frac{r_2}{R}$	$\frac{r_3}{R}$
D_0	0.4	0.4	0.4
D_1	0.8	0.2	0.2
D_2	0.8	0.3	0.1
D_3	0.9	0.2	0.1
D_4	0.6	0.35	0.25
D_5	0.55	0.45	0.25

Table 2.2 Instance 1 indices

Distribution	Index	Player 1	Player 2	Player 3
D_0	BPI	0.33	0.33	0.33
	SSPI	0.33	0.33	0.33
	HPI	0.33	0.33	0.33
	PPI	0.33	0.33	0.33
D_1	BPI	0.6	0.2	0.2
	SSPI	0.67	0.17	0.17
	HPI	0.5	0.25	0.25
	PPI	0.5	0.25	0.25
D_2	BPI	0.5	0.5	0
	SSPI	0.67	0.33	0
	HPI	0.5	0.5	0
	PPI	0.5	0.5	0
D_3	BPI	0.6	0.2	0.2
	SSPI	0.67	0.17	0.17
	HPI	0.5	0.25	0.25
	PPI	0.5	0	0.5
D_4	BPI	0.33	0.33	0.33
	SSPI	0.33	0.33	0.33
	HPI	0.33	0.33	0.33
	PPI	0.33	0.33	0.33
D_5	BPI	0.5	0.5	0
	SSPI	0.67	0.33	0
	HPI	0.5	0.5	0
	PPI	0.5	0.5	0

Table 2.3 Instance 2: four players

Distribution	$\frac{b_1}{B}$	$\frac{b_2}{B}$	$\frac{b_3}{B}$	$\frac{b_4}{B}$
D_0	0.4	0.4	0.4	0.4
D_1	0.85	0.25	0.25	0.25
D_2	0.8	0.55	0.15	0.1
D_3	0.95	0.45	0.1	0.1
D_4	0.6	0.4	0.35	0.25
D_5	0.55	0.5	0.3	0.25

three power indices are generated also in a normalized form so that they add up to one. The values of indices for these cases are shown in Tables 2.2 and 2.4.

Differences between the indices' values become more pronounced as the number of players increases (the increased number of possible coalitions allows such differences to show) (Tables 2.3 and 2.4).

Table 2.4 Instance 2 indices

Distribution	Index	Player 1	Player 2	Player 3	Player 4
D_0	BPI	0.25	0.25	0.25	0.25
	SSPI	0.25	0.25	0.25	0.25
	HPI	0.25	0.25	0.25	0.25
	PPI	0.25	0.25	0.25	0.25
D_1	BPI	0.7	0.1	0.1	0.1
	SSPI	0.75	0.083	0.083	0.083
	HPI	0.5	0.17	0.17	0.17
	PPI	0.5	0.17	0.17	0.17
D_2	BPI	0.5	0.3	0.1	0.1
	SSPI	0.75	0.17	0.042	0.042
	HPI	0.4	0.2	0.2	0.2
	PPI	0.33	0	0.33	0.33
D_3	BPI	0.7	0.1	0.1	0.1
	SSPI	0.75	0.083	0.083	0.083
	HPI	0.5	0.17	0.17	0.17
	PPI	0.5	0	0.25	0.25
D_4	BPI	0.33	0.33	0.17	0.17
	SSPI	0.33	0.33	0.17	0.17
	HPI	0.25	0.25	0.25	0.25
	PPI	0.2	0.4	0.2	0.2
D_5	BPI	0.33	0.33	0.17	0.17
	SSPI	0.33	0.33	0.17	0.17
	HPI	0.25	0.25	0.25	0.25
	PPI	0.2	0.4	0.2	0.2

2.5 Conclusion

The evaluation section has briefly demonstrated that, in general, the SSPI and BPI give similar values, favouring the players with greater available resources. A closer inspection may show that SSPI systematically does that to a slightly greater extent than BPI. The remaining indices, namely the HPI and PPI, give a higher power to relatively weaker players. This is because weaker players have smaller contributions and hence are more often found in coalitions which are minimal winning, or by-least winning for some players.

It has been shown throughout the chapter that the HPI and PPI are more appropriate indices for the coalition selection game, since they exclude a number of coalitions which are not stable and hence would not appear if nearby IoT nodes formed coalitions independently. Comparing these two indices, we can argue that the PPI is more fair, in the sense that it only considers stable coalitions in the inner core that would be formed if payoffs were allocated proportionally to players' contributions, i.e. in a fair manner.

References

1. Antoniou, J., Koukoutsidis, I., Pitsillides, A., Stavrakakis, I.: Access network synthesis in next generation networks. Comput. Netw. **53**(15), 2716–2726 (2009)
2. Banzhaf, J.: Weighted voting doesn't work: a mathematical analysis. Rutgers Law Rev. **19**(2), 317–343 (1965)
3. Brams, S.J., Fishburn, P.C.: When is size a liability? Bargaining power in minimal winning coalitions. C.V. Starr Center for Applied Economics, New York University (Working Papers) (1994)
4. Brams, S.J., Fishburn P.C.: Minimal winning coalitions in weighted-majority voting games. Soc. Choice Welf. **13**(4), 397–417 (1996)
5. Gillies, D.: Solutions to non-zero games. In: Contributions to the Theory of Games. Annals of Mathematical Studies no. 40, vol. 4, pp. 47–85. Princeton University Press, Princeton (1959)
6. Haradau, R., Napel, S.: Holler-Packel value and index: a new characterization. Homo Oeconomicus **24**(2), 255–268 (2007)
7. Microsoft Azure: Internet of Things (IoT) Bot Scenario. Azure Bot Service (2017). https://docs.microsoft.com/en-us/azure/bot-service/bot-service-scenario-internet-things?view=azure-bot-service-3.0. Accessed 15 Jan 2019
8. Myerson, R.B.: Game Theory: Analysis of Conflict. Harvard University Press, Cambridge (2004)
9. Riker, W.H.: The Theory of Political Coalitions. Yale University Press, New Haven (1962)
10. Shapley, L.S., Shubik, M.: A method for evaluating the distribution of power in a committee system. Am. Polit. Sci. Rev. **48**(3), 787–792 (1954)
11. Stone-Gross, B. et. al.: Your botnet is my botnet: analysis of a botnet takeover. In: Proceedings of the 16th ACM Conference on Computer and Communication Security, pp. 635–647 (2009)

Chapter 3
Using Game Theory to Address Mobile Data Offloading in 5G

Abstract This chapter addresses the situation where 5G communication service is enabled by data offloading from infrastructure transmission points onto the mobile users that act as content repeaters, for a monetary payoff. The aim is to enhance the communication service performance, especially in terms of traffic delays, which is a requirement in 5G networks. The offloading algorithm is viewed as an ad hoc service where the path of the traffic through the content repeating devices is selected through a negotiation process between the mobile users and the infrastructure points, as well as between the mobile users themselves. The chapter proposes a game theoretic modelling approach for modelling the strategies of the interacting nodes using a game model that resembles an auction model. Specifically, each player may take one of the two roles during an interaction, either the role of a price setter, a bidder, or the role of price taker, an auctioneer. We assume that all packets are homogeneous in the sense that they do not differ in terms of the payoff to the forwarder; therefore, the decision of whether to forward a traffic packet by a node, or to choose the node that will forward the traffic packet (depending on whether a player is currently in the role of price setter or price taker), does not depend on the traffic packet itself. The decisions are taken sequentially, i.e. the price setters advertise their prices and then the price taker selects one of them to forward the traffic packets to.

Keywords Mobile data offloading · Auctions · Mobile infrastructure · Sequential-moves strategy

3.1 Introduction

3.1.1 Overview of Technology Considered

High performance communications are targeted by 5G networks, which mainly translates in high data rates, reduced traffic delays and additional features such as energy savings and high device connectivity. As the use of big data is making its way into our everyday lives, through the increasing deployment of the Internet of things

© Springer Nature Switzerland AG 2020   43
J. Antoniou, *Game Theory, the Internet of Things and 5G Networks*,
EAI/Springer Innovations in Communication and Computing,
https://doi.org/10.1007/978-3-030-16844-5_3

Fig. 3.1 As the use of big data is making its way into our everyday lives, through the increasing deployment of Internet of things (IoT) applications, 5G networks are faced with a new challenge

(IoT), and IoT applications, 5G networks are faced with the challenge of handling the enormous growth in size of mobile data, in terms of storage and computation, coupled with an increasing need to rely on mobile devices (smart devices in their majority) to support this excess of traffic flows (Fig. 3.1).

These mobile devices carry differing characteristics and capabilities supporting the decoupling of computation and content, which can also be offloaded onto cloud storage. Such enhanced mobile user devices can even assume the roles of service or content providers, virtually extending the mobile infrastructure to include them. Furthermore, the common thread that links all this heterogeneity is the support for a user-centric paradigm of communication, converging all activities to the system's key function, i.e. to satisfy its customers. 5G networks plan to take advantage of these varying characteristics, exploiting them in complementary manners in order to achieve to surpass any limits imposed by any mobile infrastructure on its own, through appropriate synergies with the 5G network users themselves.

Synergies, i.e. cooperation between participating entities in 5G communication networks, promote the useful co-existence of heterogeneous entities, aiming at enhancing the overall network, e.g. by extending its coverage through user-initiated content repetition, since the support of demanding multimedia services, such as interactive and multiparty multimedia services (to deliver real-time and even critical time applications), becomes a challenging task due to the heterogeneity of the entities involved, the user(s) and the network infrastructure. This heterogeneity results in different and often conflicting interests for these entities. Since cooperation between these entities, if achieved, is expected to be beneficial, we pose the following question: Can cooperation be motivated in interactive situations arising in 5G communication networks, and if yes, is it beneficial for the interacting entities, especially the customers themselves, i.e. the end-user nodes? In pursue of answering this question, this chapter isolates and studies an interactive situations between user(s) and the network infrastructure points of transmission, in particular infrastructure points that belong to the mobile network, and proposes appropriate modes of behaviour that allow the interacting entities to achieve own satisfaction, in this example through the use of monetary payoff, while at the same time enhancing the performance of the 5G network.

Interactions between entities with conflicting interests follow action plans designed, by each entity, in such a way as to achieve a particular selfish goal, such interactions are known as strategic interactions. Strategic interactions are studied by game theory, which develops models that prescribe actions for entities interacting in a strategic manner, such that they achieve satisfactory gains from the situation. To target the question of how to behave in interactive situations between heterogeneous entities in 5G Networks, the chapter utilizes game theoretical models to propose models of selected strategic situations and investigates profitable behaviours of the participating entities. The study shows that cooperation can be motivated in each of the selected interactive situations and, furthermore, that such cooperative behaviour can be beneficial for the interacting entities.

3.1.2 On Mobile Data Offloading

The idea of mobile data offloading is actually the use of complementary network technologies for delivering data originally meant to be carried by the network infrastructure. Offloading reduces the amount of data being carried onto the infrastructure resources, therefore, freeing bandwidth for more of the network's users. It is also a good alternative in the case the infrastructure resources are not available, e.g. it can be used in situations where local signal reception may be poor, allowing the user to connect via alternative points and achieve better connectivity.

Mobile data offloading actions may be initiated by either an end-user (mobile subscriber) or an operator (controlling the infrastructure points). The explosion of Internet data traffic has increased the need for offloading solutions to be available in 5G, since there exists a growing portion of traffic going through mobile networks. Offloading has been enabled by smartphone devices possessing Wi-Fi capabilities since Wi-Fi is typically much less costly to build than cellular networks.

In addition to the large amounts of data, there exists a subsequent need for an increase in connection speeds in 5G networks, because of new real-time and critical time services that 5G aims to support. The need for increased 5G speeds motivates a new perspective on how the mobile user is viewed by the network, i.e. the mobile user undertakes a more active role in the network. Specifically, the mobile user is now required to participate more actively in the communication stream as a temporary infrastructure point to help carry the content through to its destination, by becoming a content repeater. In addition to alleviating congestion at the network edge, this approach may result in decreased delays.

It has been often observed that the Internet's edge, where the user devices are located, is characterized by congestion. The chapter considers the scenario where mobile user devices can be used to offload infrastructure transmission points, thus alleviating congestion at the edge of the Internet, and offering a faster solution for the path that a communication stream takes, from source to destination. This results in a seemingly denser infrastructure without actually needing to deploy new, permanent infrastructure points, but by *renting* out resources from user devices that are interested to act as potential content repeaters.

Within the definition of this scenario, the chapter explores a potential offloading strategy, based on mobile user interactions (as these are expected to take place along the path of the communication stream). The strategy is based on a game theoretic model that employs the idea of auctioning the offloading option in return for some monetary payoff offered by the infrastructure points that wish to *rent* the resources of mobile users to support a particular communication stream.

Specifically, the presented game model enables the ad hoc cooperation among mobile users along the path of the communication stream, for the purpose of allowing these mobile users to spontaneously forward some of the traffic on behalf of congested infrastructure points. The incentive for the mobile users would be the monetary payoffs offered by the operators managing the congested infrastructure points. Moreover, the operators themselves would be motivated to engage in this scenario as such agreements would significantly decrease infrastructure costs, without compromising their customer experience. This is referred to as a mobile data offloading scenario (Fig. 3.2).

An additional advantage for 5G networks is that by employing this type of scenario, a natural coverage extension is achieved, which is very important for

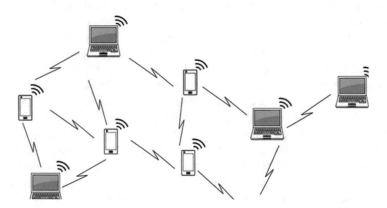

Fig. 3.2 The scenario considers the ad hoc cooperation among mobile users along the path of the communication stream, for the purpose of allowing these mobile users to spontaneously forward some of the traffic

the 5G operators. Also, the scenario offers additional redundancy in terms of communication paths that results in increased resilience of the 5G network in case of any infrastructure damage or failure.

3.2 The Offloading Scenario Elaborated

In a 5G mobile offloading scenario, we consider that each of the mobile users that are available to undertake the role of a content repeater may take one of the two game profiles during an interaction. These two profiles or roles include: the role of a price setter, and the role of price taker. Note that the *price* is initially set by the infrastructure point, motivated by the mobile network operator, that actually initiates the mobile offloading process.

Once the infrastructure point negotiates with the available user nodes and decides which node will be the node that will forward the traffic packet, the payment is made to the forwarding node, which in turn will act as the next price taker and will make subsequent payment to the selected price setter that will forward the traffic packet next, and so on. In addition to the available budget advertised by the price taker, i.e. the maximum price for which the price setters will bid, a possible fine is advertised, which will be imposed by the infrastructure, i.e. the network operator, in case the selected price setting node will not carry out its forwarding task successfully. As will be discussed later, this fine is mainly to prevent malicious behaviour by the user nodes in an attempt to gain monetary payoff without sharing their resources to support the communication stream.

One of the assumptions that can be made to simplify the model outline is that all traffic packets are homogeneous. This is not the case in terms of traffic and data characteristics in the 5G network but it can be considered in the sense that the traffic packets do not differ in terms of the payoff to the forwarding nodes; therefore, the decision of whether to forward a traffic packet or to choose the next forwarding node (whether a player is currently in the role of price setter or price taker) does not depend on the traffic packet itself, forcing the decision-making to depend solely on the interacting nodes, i.e. the potential content repeaters.

The process of creating the path through the 5G mobile user population that will act as content repeaters to facilitate the communication service is a series of decisions taken by the mobile users that are positioned in a favourable location to undertake this task, between the transmission source and destination of the specific communication stream. The decisions are taken sequentially, i.e. the price setters advertise their prices, which are then considered by the price taker, who in turn selects one of them to forward the packet to. Thus, the resulting scenario is that of a competitive situation between the price setters, since one price taker selects one of the price setters based on some criteria (mainly the advertised price), causing the price setters to be antagonistic in their attempt to be selected. Each mobile user node, also referred to as a player in the game model, may assume both of the two roles, as a traffic packet makes its way through the mobile user population in an attempt to reach each destination.

In the 5G mobile offloading game, the price setters may offer similar or the same price for forwarding a packet. In addition to the price being advertised the price taker may consider additional factors for making the final selection (such could be distance from target, overall topology, probability of a packet reaching the destination, etc.). Let's call this parameter: other additional factors (OAF) and let it represent a cost for the price taker. So, even if prices advertised are equal, we assume that OAF may still affect a price taker's decision. Note here that in the absence of OAF the price taker will always select the lowest price. Thus, the utility function of a price taker could be an additive function U_i, such that:

$$U_i = (B - k) - f(\text{OAF}), \tag{3.1}$$

where $B - k$ is the resulting budget, with B representing the original budget and k representing the selected advertised price that the price taker will have to pay to the selected price setter. A function depending on OAF represents the risk relevant to the success of the traffic packet reaching its destination, which does not depend on the price. This risk should be minimized in order to avoid possible fines imposed by the infrastructure.

The result of this utility function should be considered by the price setter in setting the advertised price. The advertised price is enticing for the price taker and competitive if $U_i \geq 0$. At the same time the advertised price should not result in a loss for the price setter. Thus, the overall payoff for the price setter could be indicated by:

$$\Pi_i = \sum_{i=1}^{x} k_i - \sum_{i=1}^{x} \phi_i, \tag{3.2}$$

for all forwarded packets where x is the total number of packets forwarded, and ϕ_i represents the fine for packet i.

Note that a fine may be imposed by the infrastructure upon mobile nodes that commit to act as content repeaters but end up not being successful in forwarding the traffic packet. This could be done maliciously, in an attempt to acquire the monetary payoff but not share any resources to facilitate the path of a traffic packet to its destination. Thus, such nodes would be punished by the imposition of fines in order to prevent similar cheating behaviour in the future.

Given that for a number of packets the player assumes the role of a price taker and pays an amount $B - k$ to a neighbour to forward a packet, then the overall payoff of the game for player i would be the sum of the monetary amount acquired to forward each traffic packet, minus the sum of the amounts that the player would have to subsequently pay to its neighbouring nodes to forward the packet further along its path. This should be taken into consideration when advertising the available budget and potential fines.

Given the existence of fines and considering that multiple mobile nodes need to cooperate along the communication path in order to achieve a successful offloading service (and thus must share the monetary payoff offered by the infrastructure points), some decisions must be made in advance by the participating mobile users. The main decision for each price setter in this situation is about selecting the price at which to advertise its own forwarding service in each interactive situation, being fully aware that simultaneously, other price setters will be advertising their own prices as well. Regarding the fine, the players know that there exists a fine, which is advertised by the price taker and it is the same for all competing mobile nodes, having no control over it. Thus, the price advertised for a single packet should consider this as fixed and evaluate the probability of the risk for the particular traffic packet to be fined as δ, resulting in an estimation for the advertised price P as:

$$P_i = k_i - \delta\,\phi_i, \tag{3.3}$$

where $Pi \geq 0$.

Another consideration towards a solution for the offloading game model would be to form groups of mobile nodes that collaborate in order to achieve higher payoff by strategically setting prices together, rather than each one node setting a price on its own. Let's consider the case where any two nearby mobile nodes decide to cooperate and set their prices strategically. Consider a possible strategy for two nodes cooperating in order to maximize their chances of winning the chance to forward a traffic packet, and subsequently sharing the payoffs:

The two interacting players, will be placed close together so they may set prices for the same packets. Initially, the players will set a price equal to the possible cost, i.e. equal to the advertised fine. Thus, if either of the two players is selected and the packet does not reach the destination, the forwarding player will not earn any profits but will not have any loss. Given that this will not happen each time, the overall payoff will be positive. In fact, at the initial bidding no other player has an incentive to set a price lower than the fine, as this is risking a negative payoff. This attitude might change during the game as the budgets of the players change, and thus in the beginning the strategy should allow the budget to remain positive. Moreover, if the players set the advertised price equal to the fine, then if the other players raise their price above the fine, then they will earn nothing, since forwarding nodes will probably select the lower price out of the advertised prices (this might not be true as the topology may also affect the decision, such as other additional factors (OAF) mentioned above). Therefore at least one of the competing mobile nodes should always advertise a competitive price, i.e. at cost/fine price.

In fact, no other price is equilibrium in such a situation. If all players set the same price above the cost and share the forwarding requests, then each player has an incentive to undercut the others by an arbitrarily small amount and capture all forwarding requests, increasing its profits. So there can be no equilibrium with all players advertising the same price above the cost, since strategically each one will aim to undercut the others. There can be no equilibrium with players advertising different prices. The players setting the higher price will earn nothing (the lower price serves all of the requests). The only equilibrium is when all players set price equal to their possible cost, which would be the set fine.

The strategy may be modified during the game in order to achieve a more competitive approach for the team of the two cooperating nodes. Since the cost is not set but may occur at a risk factor, and also we are looking at a repeated game, then players have an incentive to advertise even lower than the possible cost without reaching negative payoffs by simply monitoring their cumulative payoff and adjusting the risk by the probability that a packet will not reach its destination. This probability could be adapted throughout the game (implement an adaptive strategy based on the history of successful forwards). Thus, it would be beneficial to have a collaboration between at least two players, where one player would stay at the equilibrium strategy and the other to randomly drop its price to a value that takes the risk of paying the fine into account, so it can capture some additional forwarding requests (increase in quantity of forwards will increase the overall payoffs).

For the cooperative strategy between the two players to succeed, the players would have to eventually share the total payoff; otherwise, the player that would need to drop the advertised price below the fine threshold would not be motivated to participate in the competition.

It is noteworthy here to mention that the price taker must advertise the budget and the fine and a decision needs to be made in terms of maximum budget and maximum fine since they don't have to be the same at each hop. More specifically, at each hop the node that currently must take the role of the price taker must make a decision of what the fine should be and advertise this to the nodes that will take the role of the price setters in an attempt to win the chance of forwarding the packet and receiving the monetary payoff analogous for this action (as this is also advertised at each hop by the price taker). This must not necessarily be the same budget as in the previous hop (since the strategy is based on an ad hoc negotiation and forwarding process). Therefore, in the case of the price taker, the advertised fine should be equal or less than the advertised budget, which in turn should be equal or less to the available budget of a node since the upstream node must immediately pay the selected downstream node, i.e. the price taker must immediately pay the price setter once it is selected. Finally, as briefly mentioned earlier, the price taker needs a strategy for selecting the forwarding node, which could be based only on advertised price but also on other factors such as the topology, etc.

3.3 Single User Strategies for Resolving the Offloading Scenario Using an Auction Model Approach

3.3.1 Auctioning Strategy

Consider the topology illustrated in Fig. 3.3. In this section we consider strategies for single players that are not additive functions as the abovementioned scenarios

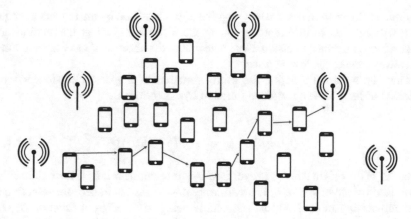

Fig. 3.3 A set of infrastructure points is complemented by the use of resources from mobile users that may support through the offloading service, a path of the communication service through ad hoc collaborations between them

and we try to quantify a sequence of actions that characterize an interaction based on the idea of auctions, where the traffic packet is *auctioned*, and the nearby mobile users may *bid* for the chance to forward it and, thus, receive a monetary reward.

For this section, we refer to the *price taker* mobile node as the *auctioneer*, whereas we refer to the *price setter* mobile node as the *bidder*, in order to keep in line with the general philosophy of the auction game model. One of the main strategic moves of the price taker or the auctioneer is that at all times, the auctioneer will set a marginal decrease in the advertised budget compared to that of the previous hop. Moreover, the fine will remain equal to that of the previous hop. There is no need to decrease the fine, since the threat of punishment will only motivate more stable collaborations as the risk of losing monetary payoff increases.

In regard to downstream node selection, the auctioneer will be looking not only at the *budget-fine model* but also at additional information about the bidder that accumulate to the risk of forwarding the packet successfully. The risk may be represented as a coefficient between 0 and 1 that is adjusted according to the bidders' profiles in relation to the current forwarding request, e.g. if the bidder has high probability of successfully forwarding the traffic packet, the coefficient will be closer to 1, whereas anything less shows a problematic profile for the current forwarding request.

Considering only the *budget-fine* model, it will be highly likely that bidders will be offering the same or similar price tags for forwarding a packet. Inevitably, auctioneers should be considering additional information to be in a position to better assess candidate bidders and their potential to successfully deliver a packet to the destination. As such, the final selection could be based on factors related to the network and its topology (such could be distance from target, overall topology, probability of a packet reaching the destination, etc.). Let's call these auxiliary parameters: *other additional factors* (OAF), and let OAF reflect a utility for the auctioneer. So, even in cases when all price tags advertised by the bidders are equal, OAF will be used to differentiate among the bidders and allow the auctioneer to make a more informed decision. Note here that in the absence of OAF the auctioneer will always select the lowest price.

Thus, let's formally define the utility function of a specific auctioneer i to a specific bidder j to be a weighted function U_{ij} as follows:

$$U_{ij} = B_i - k_j - c_i \times (f(\text{OAF}_j)), \tag{3.4}$$

where B_i is the original budget advertised by the auctioneer i, k_j is the bidding price advertised by the bidder j, c_i is an arbitrary cost at the auctioneer for selecting the next forwarding node j and its magnitude being affected by a function of OAF specific to node j that indicates the risk probability for selecting j, $f(\text{OAF}_j)$. Clearly this cost is subtracted from i's budget minus the bid set by bidder j in order to compute a true utility value for auctioneer i when interacting with bidder j. The advertised price is enticing for the auctioneer and competitive if $U_{ij} \geq 0$.

To successfully forward packets to the destination, the auctioneer will be simply adjusting $f(\text{OAF}_j)$.

3.3.2   Bidding Strategy

At the same time, this utility should be considered by the bidder in setting the advertised price if it is to be competitive and win the auction. Equally though, the advertised price should not leave the bidder at a loss. Hence, the decision for each bidder in this situation is about selecting the price to advertise in each interactive situation, being fully aware that simultaneously, other bidders will be advertising their own prices and that the auctioneer behaves as detailed in (3.4). Notably, the advertised fine send out by the auctioneer is the same for all competitors that have no control over it. Thus, the price advertised for a single packet should consider this as fixed and evaluate the risk for the particular packet p using OAF.

3.3.2.1   Equilibrium Strategy

Noticeably, bidding at the advertised fine stems as the safest approach, especially when each bidder acts alone (contrary to the cooperation strategy elaborated earlier in the chapter). Actually, no other price will stabilize to equilibrium in such a situation. Indicatively, if all players set the same price above the fine and share the forwarding requests, then each player has an incentive to undercut the others by an arbitrarily small amount and capture all forwarding requests, increasing its packet delivery ratio. So there can be no equilibrium with all players advertising the same price above the fine, since strategically each one will aim to undercut the others. Moreover, there can also be no equilibrium with players advertising different prices. The bidders setting the highest price tags will risk being left out. The only equilibrium is when all players set their price tags equal to the fine.

3.3.2.2   Aggressive Strategy

On the other hand, it is well understood that each and every bid has an associated risk. This risk might be due to the strategies followed by each player, in addition to a number of network related factors (most prominently their physical location within the network topology). Moreover, since we are also looking at a repeated game, then players might have an incentive to bid at price tags lower than the fine. Since nodes might be in possession of a high balance, such an aggressive bidding that might result in negative temporary payoff would not result in an overall negative balance. Although this type of aggressive strategy has been previously discussed as part of a collaborative approach, it is not unlikely that in an attempt to increase its cumulative payoff, a node may risk to bid below the equilibrium threshold, singularly. It is however important that when such strategies are employed by singular players that the players themselves are able to monitor their cumulative payoff and accordingly adjust price tags based on tolerable risk (the risk being the probability that a packet will not reach its destination at the end). This risk factor should be dynamically adapted throughout the game, and could be implemented as an adaptive coefficient that acts as a controller for the decision to be made whether or not it is worth taking the risk for a particular interaction and packet.

3.4   An Example of a Single Selection on the Communication Path

Referring to 5G networks, we need to consider that we are dealing with a multi-level system, where decisions can be taken at different levels and by different system entities. The main task considered in this example is a single step in the process of achieving a successful chain of decision-making such that the system achieves the means to efficiently provide the requested service, and consequently, the requested communication stream from source to destination. In the specific example, we consider the decision-making to be the first step in this decision-making chain, where the infrastructure point advertises a budget and a fine and requests bids from mobile users, from which, a bidder is then selected to repeat the content, i.e. the traffic packet. Therefore, the decision-making entities are: *(i)* the mobile user bidding on the specific traffic packet and *(ii)* the infrastructure point from the operator's network acting as a service provider (*where the service is an auction that will result in the selection of a mobile user to act as a content repeater*).

Both decision-making entities are driven by "satisfaction–quantification" utilities, which are based on the individual entity's criteria. The criterion for the selection by the auctioneer of the content repeater is mainly the mobile user's low *price* in addition to a low risk function in terms of the potential for communication success.

In the case of a 5G network, the increasing availability of resources at the user device allows for multiple user devices to be able to bid for the opportunity to become content repeaters for the infrastructure point and receive appropriate monetary payoff, e.g. by their network operator. Consequently, multiple mobile users in this example may interact with the auctioneer, i.e. the infrastructure point looking to offload a traffic packet, in an attempt to achieve participation in the offloading service and consequently revenue maximization. The infrastructure point is now faced with a set of available mobile user nodes from which it may *choose* the best one according to its own utility function. The infrastructure point will select a mobile user to become a content repeater, only in the case that such a decision results in a positive utility for the specific infrastructure point.

The chapter recognizes that in a 5G network there exists the need to consider an interaction between the infrastructure and the mobile user, since this can be necessary to alleviate the increasing congestion at the edge of the network as well as for decreasing expected traffic delays and supporting real-time and critical time services. Since the proposed offloading mechanism controls the decision of which mobile user will be selected as the next content repeater, in a chain of decisions from source to destination, we refer to the following example model as the *node selection* game.

The selection mechanism decides which is the best mobile user to handle a specific step in the communication stream. The selection may be based on the user's bidding price but also on additional factors that may pose a risk to the success of the communication, such as the mobile node's context and location, in relation to any network constraints. In the case the initial conditions are compromised by dynamic changes such as the user's mobility, the node selection mechanism may be triggered so that a different, more appropriate node handles the remaining of the session (as it is often the case that more than one traffic packet will be repeated by the same selected node).

A well-designed node selection scheme can support capacity planning on a day-to-day basis by allowing the constituent mobile nodes and consequently the 5G infrastructure to make the best use of the available resources.

Once the set of available nodes is known by the infrastructure point, it may indicate its preferred node using a utility-based approach, i.e. selecting the network that appears to be the most *satisfying* choice. This preference is used as an input for subsequent packets but other criteria to be considered in subsequent decision-making rounds may include capacity availability and load balancing. Once the node is selected, the node's available resources need to be updated so that the next service request can be handled correctly.

Selecting the node to serve a traffic request that satisfies both the node and the infrastructure point may become a challenging task. A careful examination of the interaction between the two entities during the node selection mechanism is taken into consideration when the example node selection model was defined.

3.4.1   A Closer Look at the Node Interaction

The interaction between a node and an infrastructure point in a 5G network may be viewed as a negotiation between the two entities. The infrastructure point gives some kind of *compensation* (most likely monetary) and the node gives a *promise*, in this case of a specific action supporting the communication service. We will treat this interaction as a *one-shot* interaction, i.e. the two entities interact once at session activation and the decision taken is considered valid for the whole duration of the session; also, the subsequent decisions are considered independently of this decision. This is the case that the infrastructure point does not have the intelligence to *remember* a node's past behaviour, although that would significantly affect the decision-making. Therefore for this example, we maintain the assumption that no memory is kept once a decision has been made about the interaction between any node and an infrastructure point.

Investigating the interaction between the two entities, we seek to discover the incentives for each entity to select certain strategies, i.e. sets of actions. The incentive is usually realized by the payoff of each entity involved, resulting from the actions taken by each entity. In the specific interaction between the mobile node and the infrastructure point:

- the payoff of the infrastructure point acting as the auctioneer is the satisfaction level perceived by the decision, based on a utility function that considers both the bidding price and the risk generated by a function of OAF (other additional factors), where the measurements are subjectively generated by the auctioneer, and
- the payoff of the node is basically a profit made from the compensation given by the infrastructure point for the specific content combined with the probability of being fined in case the communication is not successful.

The auctioneer cannot enforce that the node satisfies the communication request but can punish the node for any action that would result in an unsuccessful communication path. The bidder is aware of the risk of being fined and the specific fine amount prior to the bidding.

In a node selection scenario, the two participating entities make sequential decisions. In particular, the auctioneer makes a decision first; he chooses a payment and a fine to advertise to the node, in order to get the desired content repetition support. Then, it is the turn of the node to make a decision of whether to bid or not for a portion of the advertised payment, considering the risk of being fined.

We utilize a notion from game theory to model this sequential decision-making: sequential-moves games. Regarding sequential-moves games, an easy way to visualize them is by illustrating the game using tree diagrams made from nodes and branches (a.k.a. game trees); they are joined decision trees for all of the players in the game, illustrating all of the possible actions that can be taken by all of the players and indicating all of the possible outcomes of the game.

According to this modelling, the auctioneer is a *first-mover* in this game; he chooses an incentive payment scheme to encourage the node to bid on the specific content. This interaction model is referred to as the *principal–agent* interaction model in game theory [1], which is also adopted in this work, with the auctioneer taking the role of the *principal* and the node taking the role of the *agent* such that the agent takes an action that affects the payoff to the principal. To further elaborate, the auctioneer offers an incentive payment as a contract to the node and requests a specific service in return.

On the other hand, the node, based on the offer provided by the auctioneer and the requested service, decides whether to bid for the offer or not. The node's decision is based on the expected payment compared to the node's subjective cost of performing the action, i.e. the risk of being fined.

If the node does not bid for the offer, the interaction terminates. If the network bids for the offer, it must provide the necessary resources to satisfy it; simultaneously, the auctioneer predicts the risk of success of the communication service as it relates to the specific node. To make the problem tractable, we examine only the possibility that an auctioneer predicts one of the two possible levels of success: *basic* quality and *high* predicted success. The payoff is adjusted accordingly, differentiated between a predicted basic or high success. Examining further this payoff exchange, we observe that the payment scheme from the auctioneer to the bidder is contractible, i.e. the payment for each different service instance may be agreed in advance. However, the node's eventual success is not contractible because in an IP-based network, quality cannot be guaranteed; there is always, no matter how small, a non-zero probability of quality degradation, and eventually unsuccessful communication.

The level of connection quality may be mapped onto the instantaneous amount of both the auctioneer's and the bidder's *satisfaction*. We model this in our one-shot game by the payoff function of the auctioneer. The satisfaction of the auctioneer in a one-shot game can be defined to be a function of requested success level and of payment offered to the network. Given requested success level q and the payment offered P, the user payoff function can be defined as follows:

$$u(q, i, P) = f(q, i)/P, \tag{3.5}$$

where $f(q, i)$ is an increasing function, and $f(q, i) \geq 0$. This function represents the prediction of the success that a bidder will have with a specific communication request. Note that this may be based on several factors indicated as OAF in previous sections, related to node i for requested success level q. When $f(q, i) = 0$ a prediction of no expected success is indicated. Positive values of $f(q, i)$ indicate that some positive expected success is predicted. Higher values of $f(q, i)$ indicate higher perceived qualities. The function $f(q, i)$ depends also on the specific mobile node under consideration, i. In order to specify the auctioneer's expected success level per unit payment offered, we divide $f(q, i)$ by P, and this gives the payoff of the auctioneer in this example of the node selection game.

The satisfaction function of the node depends on the success q expected by the auctioneer, which is mapped to an amount of resources required for service support using function $g(q)$. Given that the node has a cost function $c(g(q))$ corresponding to the cost of a network for supporting success level q, the node payoff function may be defined to be:

$$v(q, P) = \begin{cases} P - c(g(q)), & (T - g(q)) > 0 \\ 0, & (T - g(q)) \leq 0, \end{cases} \tag{3.6}$$

where T is the total amount of node resources currently available. As the payment increases, the node's payoff increases, but the higher the cost of a network to support a specific success level, the less it is *satisfied*, because this implies that the profit made per unit of used node resources decreases. Therefore, the node seeks the maximum amount of profit that it can make for a specific success level requested. So, in the case where $P - c(g(q)) \leq 0$, the node will not bid to support the specific communication request.

Game models are based on the element of rationality; however, one may argue that rationality implies that the players are perfect calculators and flawless followers of their best strategies, which is not always a correct replicate of a particular situation, thus we may assert that this is not the case. Therefore, rationality may be better described to be the players' knowledge of their own interests based on each player's own value system. Based on this element of rationality the players calculate their strategies. Strategies may consist of single actions (one-shot games) or sequences of actions (repeated interactions) and each strategy gives a complete plan of action, considering also reactions to actions that may be taken by the opponent. Strategies may be pure, i.e. provide complete definitions of how a player will play in the game (his moves), or mixed, i.e. assignments of a probability to each pure strategy. Our model employs single-action and pure strategies.

Games are motivated by *profitable* outcomes that await the players once the actions are taken. These outcomes are referred to as payoffs. Payoffs for a particular player capture everything in the outcomes that the particular player cares about. If a player faces a random prospect of outcomes, then the number associated with this prospect is the average of the payoffs associated with each component outcome, weighted by their probabilities. Since we refer to a single-action game, the calculation of payoffs should be straightforward here.

Nevertheless, an element of probability does exist in our game as well. Even though a node may promise to offer a certain level of success, this may be degraded unexpectedly because of node or infrastructure dynamic changes, such as signal strength degradation or mobility. This probable behaviour must be taken into account when payoffs are calculated for the node. If no degradation is observed, then the node's payoff is equivalent to the node payoff function $v(q, P)$.

In case of actual service degradation the auctioneer penalizes the bidder by offering a decreased amount of compensation, $v'(q, P) = a \cdot v(q, P)$, where a represents a probability of not experiencing degradation of service success; if $a = 1$, the payoff received by the node is exactly its profit, whereas if $0 < a < 1$, the payoff received by the node is decreased by a factor of a. Thus, the payoff function of the node considering the possibility of degradation becomes:

$$v'(q, P) = \begin{cases} a \cdot [P - c(g(q))], & (T - g(q)) > 0 \\ 0, & (T - g(q)) \leq 0 . \end{cases} \quad (3.7)$$

This refers to the concept of the fine introduced in the previous sections in a more simplified manner.

From the auctioneer's side, in case of degradation the payoff of the auctioneer will be affected in a similar manner, i.e. $u'(q, i, P) = a \cdot [f(q, i)/P]$. Any interaction model (i.e. any sequential-move game model) may be represented by a *tree diagram*, where the root of the tree represents the *first-mover* and the leaves of the tree give the payoffs to all interacting entities (first the payoff to the *first-mover*, then the payoff to the *second-mover*, etc.). The intermediary nodes and branches show the game flow in terms of states and decisions in an event-sequential manner.

In the given interaction model the auctioneer gives a payment to the node to support a session but since there can be two levels of expected success (*basic, high*) and corresponding levels of perceived success by the auctioneer q_L and q_H, there can be two levels of payment to the node, P_L and P_H where $P_L < P_H$. Thus, the advertised payment depends on the level of success that the auctioneer predicts for the specific communication service.

3.4.2   Bayesian Models and Truthfulness

Since one of the aspects that must be considered in this game model example is the behaviour of the participating players in terms of whether they will employ honest or dishonest strategies, we will discuss briefly in this section the idea of Bayesian games and truthfulness. Bayesian game models are useful because they allow us to consider the different types of players ahead of the game, e.g. consider a truthful type or consider a dishonest type of player. A Bayesian game [1] is a strategic form game with incomplete information attempting to model a player's knowledge of private information, such as privately observed costs, that the other player does not know. Therefore, in a Bayesian game, each player may have several types of behaviour (with a probability of behaving according to one of these types during the game). We use the Bayesian form game, in order to investigate the outcomes of an otherwise normal form game, given that each player does not know whether the opponent is truthful or dishonest.

Consider a normal form game where the players have some random payoffs. Let each player in the *normal form* game have two types: the *truthful* type and the *dishonest* type. Suppose that each of the two players has incomplete information about the other player, i.e. does not know the other player's type. Furthermore, each of the two players assigns a probability to each of the opponent's types according to own beliefs and evaluations. Let p_i^l be the probability according to which player i believes that the opponent is likely to be of type *truthful*, and $p_i^h = (1 - p_i^l)$ be the probability according to which player i believes that the opponent is likely to be of type *dishonest*.

Since the two players are identical, i.e. they have the same two types and the same choice of two actions, we will only analyse player i; conclusions also hold for player j, where $i, j \in [2], i \neq j$. Therefore, player i believes that player j is of type *truthful* with probability p_i^l, and of type *dishonest* with probability $1 - p_i^l$. Each player has a choice between two possible actions: to admit its type or to lie. Bayesian games show that if player i believes that the probability p_i^l, i.e. that player j is of type *truthful*, is higher than the probability p_i^h, then it is more motivated to lie, where $i, j \in [2], i \neq j$. If player i believes that the probability p_i^h, i.e. that player j is of type *dishonest*, is higher than the probability p_i^l, then it is more motivated to lie, where $i, j \in [2], i \neq j$. Therefore, by the above statements the two players playing the Bayesian form of a *normal form* game are not motivated to be truthful but instead they are motivated to cheat in order to get greater payoffs. This is shown to be true in a normal form game model with Bayesian players [1].

In order to motivate the two players to be truthful, there must exist a mechanism that can penalize a player who turns out to lie, assuming that it is detectable whether a player has lied or not. We refer to such mechanisms as pricing mechanisms [3]. Let the lies be detectable after a game round has been complete, whether either of the participating players has lied. In order to motivate the players to be truthful, we may introduce a *pricing mechanism*, i.e. a new variable that tunes the resulting payoffs, in the payoff function of each player. A side effect for a player that decides to cheat is that it risks not to receive its expected payoff as the dishonest behaviour may cause future payoff discounts. The pricing mechanism is a post-game punishment, i.e. cheating in a game does not affect the game round in which a player cheats but subsequent games. Thus, a state of history of a player's behaviour in similar interactions must be kept.

3.5 Conclusion

A final step to this work would consist of an implementation of the model, with a selection of numerical values for the payoffs and a randomly generated probability of success. The implementation could simply make use of a custom numerical simulator. The desired outputs from the simulation of single node selection game, such as the one described in the previous section, would be the accumulated node payoffs after multiple repetitions of the game, i.e. multiple interactions with the infrastructure points or with other user nodes, following the above-mentioned rules at each selection round and eventually summing up all the payoffs from all the rounds. This can be an interesting set of simulations, in order to determine whether the nodes are indeed satisfied with the auctioneer-based selection decision (*since the model maximizes the auctioneer's payoff*). Thus the aim of implementing and running a simulation model would be to show how the payoffs behave after an increasing number of iterations.

Since the required outputs are the player payoffs, the element of time is not an important factor in such a simulation model. The importance of the time factor in a simulation indicates whether there is a need for a static or a dynamic simulation. Given the time requirements for the simulation model of single node selection, one may implement a static simulation model, without compromising the results. A second factor that must consider is that the selection decision itself, as well as the node payoffs, are based on parameters, which are not deterministic, i.e. which cannot be determined in advance. This implies that some of the system variables must be random, and in turn the simulation model must support random (*a.k.a stochastic*) evaluation. Hence, the simulation model must be a stochastic, static model (*such models are known as Monte Carlo models* [2]). The implementation of a Monte Carlo simulation model is out of the scope of this chapter.

References

1. Gintis, H.: Game Theory Evolving: A Problem-Centered Introduction to Modeling Strategic Interaction. Princeton University Press, Princeton (2000)
2. Mooney, C.Z.: Monte Carlo Simulation. Quantitative Applications in the Social Sciences. Sage Publications, Thousand Oaks (1997)
3. Nisan, N., Ronen, A.: Algorithmic mechanism design. Games Econom. Behav. **35**(1–2), 166–196 (2001)

Chapter 4
Using Game Theory to Motivate Trust in Ad Hoc Vehicular Networks

Abstract This chapter considers the cooperation between vehicular nodes in an ad hoc network as a decision resulting from the potential interaction between any two such nodes. We consider the potential for continuous communication between these nodes, under the condition they keep sharing mutually beneficial information. Describing and analysing entity interactions is a situation that makes a good candidate to be modelled using the theoretical framework of game theory. The fact that this new and interesting mode of human interaction between vehicles does not need to be a human-initiated process, but a process that can be initiated by an object-to-object communication, because of IoT, brings up the issue of establishing a trusted communication between the interacting parties. However, establishing such trusted, cooperative behaviour in vehicular networks is not an easy task, even if vehicular networks are considered to be user-centric networks. Game theory provides appropriate models and tools to handle multiple, interacting entities attempting to make decisions and seeking a strategic solution state that maximizes each entity's utility, incorporating a consideration of trust within that utility. Game theory has been extensively used in networking research as a theoretical decision-making framework, and this chapter makes use of this know-how.

Keywords Vehicular networks · Game theory · Prisoner's dilemma · Trust model

4.1 Introduction of a Vehicular Node Interaction Scenario

5G networks aim to support communication of vehicles on the road, so that moving vehicles can form ad hoc networks in order to share or exchange useful and interesting information. However, the ad hoc nature of these networks and the fact that sharing information would imply that a vehicular device would have to share resources with another such device, which may not be trustworthy, have raised many issues related to trust and security. These issues need to be addressed prior to deploying such services in a 5G vehicular network.

© Springer Nature Switzerland AG 2020   61
J. Antoniou, *Game Theory, the Internet of Things and 5G Networks*,
EAI/Springer Innovations in Communication and Computing,
https://doi.org/10.1007/978-3-030-16844-5_4

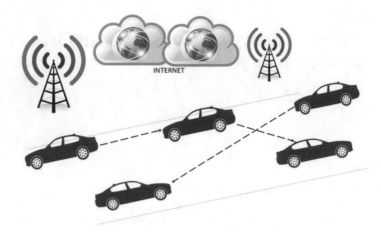

Fig. 4.1 An illustration of a vehicular network, highlighting the communication between the vehicles as these may form ad hoc connections according to their specific travel location

To rectify the aforementioned issues, this chapter approaches the cooperation (i.e. trusted relationship) between two vehicles i and j in an ad hoc vehicular network, as a decision resulting from the potential interactions between the vehicular nodes. These interactions could be finite (i.e. exchanging few messages during a short time period), or infinite (i.e. i and j keep sharing mutual beneficial information). It is worth noting that interactions between vehicles are often repeated and there is no knowledge of the number of such interactions that can take place over time. The repeated nature of the interaction comes from the premise that 5G is mainly deployed in dense urban areas, and thus, vehicles that cross each other's paths during certain times, e.g. morning rush hour, are bound to cross each other's path again if they keep to the same driving routine (Fig. 4.1).

It is important to note that these interactions often incur a cost for both sides. For the sender, sharing quality information (e.g. traffic related) with the receiver incurs a cost that includes calculations, communications and the risk of being exposed to a stranger on the road. For the receiver, the risk of trusting the received information to act upon and returning the favour to the sender, who is also a stranger, is the main cost.

Initially, there is no motivation for either vehicular node travelling on the road to cooperate, i.e. to share information or to accept information from other vehicular nodes. Hence, this chapter aims to demonstrate how such *trust* could be mathematically motivated, and hence a cooperative attitude among vehicular nodes can also be motivated, in order to eventually establish trusted relationships that bring mutual benefits (i.e. payoffs) for both sides. Consequently, nurturing the established relationship for continuous communication under the condition they keep sharing mutually beneficial trusted information.

To achieve the paper's aim and incentivise vehicular nodes to share valuable information over the ad hoc vehicular networks, a trust model should be in place

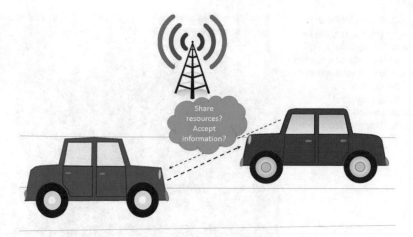

Fig. 4.2 An illustration of a two vehicular nodes travelling along the same road with an opportunity to share information. However, motivation is necessary in order for such nodes to be able to trust each other in a vehicular network

to promote cooperation and trusted relationships in these ad hoc networks. This model should highlight the mutual benefits for cooperative travellers and possible losses of a non-cooperative behaviour (i.e. defection or cheating). Describing and analysing travelling node interactions in ad hoc vehicular networks is a situation that makes it a good candidate to be modelled using the theoretical framework of game theory. Game theory provides appropriate models and tools to handle multiple, interacting entities attempting to make decisions and seeking a strategic solution state that maximizes each node's benefit (Fig. 4.2).

The proposed approach proves that adopting cooperation strategies while interacting with others in vehicular networks results in (1) the highest payoffs (i.e. benefits), and (2) in an increase in trust between the cooperating users (who gain more social interactions and better reputation). It is shown that the developed approach can establish direct and indirect trusted relationships in these networks showing also mathematically the viability and the efficiency of the proposed approach.

Game theory has been extensively used in networking research as a theoretical decision-making framework. In order for a strategic situation to become a *game* between two or more players, there must be a mutual awareness of the participants regarding the cross-effect of their actions. A strategic situation, where the actions of a participant may alter another's outcome, is primarily characterized by the players' strategies. In addition, a strategic situation contains other elements that must be taken into consideration when modelling such a situation as a game, e.g. chance and skill (elements that are not easily controlled or modified). Although game models employ the element of rationality, one may argue that rationality implies that the players are perfect calculators and flawless followers of their best strategies, which is not always a correct replicate of a particular situation, thus we may define

Fig. 4.3 The solution to a
strategic game is derived by
establishing equilibria, i.e.
finding a solution of a specific
combination of strategies
such that none of the
interacting players is
motivated to modify the
selected strategy combination

rationality to be the players' knowledge of their own interests based on each player's
own value system. Based on this element of rationality the players calculate their
possible strategies.

The solution to a strategic game is derived by establishing equilibria. Equilibria
may be reached during the interaction of players' strategies when each player is
using the strategy that is the best response to the strategies of the other players (i.e.
given the strategies of the other players, the selected strategy results in the highest
payoffs for each player participating in the game). The idea of equilibrium is a useful
descriptive tool and furthermore, an effective organizing concept for analysing a
game theoretic model (Fig. 4.3).

The *prisoner's dilemma* and *iterated prisoner's dilemma* (IPD) have been a rich
source of research material since the 1950s. However, the publication of Axelrod's
book in 1984 [1] was the main reason that this research was brought to the attention
of other areas outside of game theory, as a model for promoting cooperation. In fact,
IPD is now fundamental to certain theories of cooperation and trust. Subsequent
editions of Axelrod's *The Evolution of Cooperation* feature a quote by the Wall
Street Journal that mentioned in reference to the book (see Fig. 4.4): *"Our ideas of
cooperation will never be the same"*.

4.2   A Game Theoretic Vehicular Node Interaction Model

In order for a strategic situation to become a *game* between two or more players,
there must be a mutual awareness of the participants regarding the cross-effect
of their actions. A strategic situation, where the actions of a participant may alter
another's outcome, is primarily characterized by the players' strategies.

"Our ideas of cooperation will never be the same."
— *The Wall Street Journal*

REVISED EDITION

THE EVOLUTION OF

COOPERATION

Robert Axelrod

With a new Foreword by Richard Dawkins

Fig. 4.4 Axelrod's book was a milestone publication for game theory

In addition, a strategic situation contains other elements that must be taken into consideration when modelling such a situation as a game, e.g. chance and skill, (elements that are not easily controlled or modified). Game models, i.e. models of specific strategic situations, may be categorized in various ways due to the several elements that they contain. A usual categorization is made by looking at the players' movements; if they are sequential we have an extensive game form (a.k.a. sequential-moves game form), whereas if they are simultaneous the game form is referred to as normal. Furthermore, an interaction may happen only once or repeatedly; in the first situation we are faced with one-shot game models, while the second situation requires repeated game models.

An additional dichotomy is whether the players are in complete conflict, where the model employed is a non-cooperative one, or they have some commonality, where a more cooperative game model may be more appropriate. Finally, another important categorization is whether we are dealing with a game where the players have complete information about all actions taken or only partial information. When characterizing a game it is important to keep in mind the various possible categorizations of game models in order to better describe the required strategic situation as completely as possible.

Next, we take a look at the prisoner's dilemma and iterated prisoner's dilemma (IPD) as solution models for the vehicular node interaction. The prisoner's dilemma is basically a model of a game, where two players must decide whether to cooperate with their opponent or whether to defect from cooperation. Both players make a decision without knowing the decision of their opponent, and only after the individual decisions are made, these are revealed. The desirable cooperative behaviour must be somehow motivated so that the players' selfish but rational reasoning results in the cooperative decision. Cooperation may evolve from playing the game repeatedly, against the same opponent. This is referred to as iterated prisoner's dilemma, which is based on a repeated game model with an *unknown* or infinite number of repetitions, also referred to as game horizon. The decisions at such games, which are taken at each repetition of the game, are affected by past actions and future expectations, resulting in strategies that motivate cooperation.

We consider that node interaction in a vehicular network may be modelled by an IPD. We show that our proposed model for vehicular node interaction falls within the prisoner's dilemma, and specifically, the IPD category for strategic interactions and, subsequently, we provide proof that cooperation is an equilibrium behaviour for the outlined game model.

The vehicular node interaction model is defined as follows.

Definition 4.1 Consider the situation: Let a receiving vehicular node i have an internal evaluation of the significance level q of received information, by a particular sending vehicular node $j \neq i$. The expected benefit from receiving information for node i is thus $\pi(q)$, a proportionally increasing function of the significance of the received information, e.g. increased traffic would score a lower significance than a car crash or road constructions. The cost of interaction is modelled as a constant ρ, which represents the risk in terms of resources that i needs to share

Table 4.1 Receiving and sending node interaction as a prisoner's dilemma game

	Send. node cooperates	Send. node cheats
Rec. node cooperates	$\pi(q) - \rho, \rho - \tau(q)$	$\pi(q') - \rho, \rho - \tau(q')$
Rec. node cheats	$\pi(q) - \rho', \rho' - \tau(q)$	$\pi(q') - \rho', \rho' - \tau(q')$

with j. Therefore, the utility of a receiving node i, $U_i = \pi(q) - \rho$. On the other hand, ρ is a positive payoff for the sending node, i.e. a representation of additional resources received through i. The sending node also bears certain cost, for sending information of significance level q. Let's define this as $\tau(q)$. Thus, j has a utility function $U_j = \rho - \tau(q)$.

If the sending node decides to send malicious or incomplete information to save resources, its utility function may generate a higher payoff in a single interaction, i.e. $U_j = \rho - \tau(q') > U_j = \rho - \tau(q)$.

If the receiving node decides to cheat and not share the necessary amount of resources with the sending node, its utility function may also generate a higher payoff in a single interaction, i.e. $U_i = \pi(q) - \rho' > U_i = \pi(q) - \rho$.

This situation is referred to as the user–network interaction game.

Let's consider a single interaction to show how the communication falls under the prisoner's dilemma interaction model.

In Table 4.1 we define the players' utility functions, where ρ represents the cost for the receiving node, but also the amount of received resources as a gain for the sending node, $\pi(q)$ represents the benefit resulting from the received information for the receiving node, and $\tau(q)$ represents the resource cost for sending information for the sending node, with q representing the significance level of this information. Note that the action of defecting in a prisoner's dilemma is represented in the case of the vehicular node interaction game by the action of *cheating*, i.e. by sending malicious or inaccurate information by the sending node or not sharing sufficient resources by the receiving node.

Proposition 4.1 *Consider a single interaction of the vehicular node interaction game. Then the game is equivalent to a prisoner's dilemma game.*

Proof Our proof utilizes the following:

Definition 4.2 Consider n one-shot game with two players in which the player has two possible actions: to cooperate with his opponent or to cheat. Furthermore, assume that the two following additional restrictions on the payoffs are satisfied [4]:

1. The order of the payoffs for each player is such that $A > B > C > D$, where A is the payoff for cheating when opponent cooperates, B is the payoff when both cooperate, C is the payoff when both cheat and D is the payoff for cooperating when opponent is cheating.

Table 4.2 Sending node and
receiving node payoffs
ordered according to
A, B, C, D

	Rec. node payoffs	Send. node payoffs
A	$\pi(q) - \rho'$	$\rho - \tau(q')$
B	$\pi(q) - \rho$	$\rho - \tau(q)$
C	$\pi(q') - \rho'$	$\rho' - \tau(q')$
D	$\pi(q') - \rho$	$\rho' - \tau(q)$

2. The reward for mutual cooperation should be such that each player is not
   motivated to exploit his opponent with probability 50% and be exploited with
   the same probability.

Then, we say that the game is equivalent to the prisoner's dilemma game.

Claim In the vehicular node interaction game there are two possible actions for the
two players: to cooperate and to cheat.

Therefore, the actions of the players in the vehicular node interaction game
match the actions of the players of a prisoner's dilemma game. Table 4.2 maps
each player's payoffs, found in Table 4.1 to actions A, B, C, D as defined in
Definition 4.2.

Lemma 4.1 *For both the receiving node and the sending it holds that:* $A > B >$
$C > D$.

Proof Examining the receiving node, we verify straightforward that $\pi(q) - \rho' >$
$\pi(q) - \rho$, thus $A > B$, and that $\pi(q') - \rho' > \pi(q') - \rho$, thus $C > D$, since $\rho > \rho'$.
It must also hold that $\pi(q) - \rho > \pi(q') - \rho'$. Let the difference between the
maximum and minimum payoff that can be perceived be greater than the difference
between the maximum and minimum amount of resources that may be offered by
the receiving node, such that $\pi(q) - \pi(q') > \rho - \rho'$. Therefore, $\pi(q) - \rho >$
$\pi(q') - \rho'$, and $B > C$. Examining the sending node, we verify straightforward that
$\rho - \tau(q') > \rho - \tau(q)$, thus $A > B$, and that $\rho' - \tau(q') > \rho' - \tau(q)$, since $\tau(q) >$
$\tau(q')$. It must also hold that $\kappa - \tau(q) > \kappa' - \tau(q')$. Let the difference between the
maximum and minimum amount of resources that may be offered by the receiving
node be greater than the difference between the maximum and the minimum cost
undertaken by the sending node to support the corresponding significance levels,
such that $\rho - \rho' > \tau(q) - \tau(q')$. Therefore, $\rho - \tau(q) > \rho' - \tau(q')$, and $B > C$.

Lemma 4.2 *In the vehicular node interaction game, neither of the two players
plays the game in such a way as to end up being exploited with a probability of
50% and exploiting the other player with the same probability.*

Proof In fact, Lemma 4.2 implies that the reward for cooperation is greater than the
payoff for the described situation, i.e. for each player it must hold that $B > \frac{A+D}{2}$.
For the receiving node, substituting the corresponding payoffs in the expression we
get, $\pi(q) - \rho > \frac{\pi(q) - \rho' + \pi(q') - \rho}{2}$. Rearranging, we get $\pi(q) - \rho > \frac{\pi(q) + \pi(q')}{2} - \frac{\rho' + \rho}{2}$,
from which the condition holds since $\pi(q) > \pi(q')$ and $\rho > \rho'$. For the sending

Fig. 4.5 The classic prisoner's dilemma game describes the situation where two prisoners, arrested for the same crime, are interrogated independently by the police officers, given incentives to defect against their partner. Each prisoner must decide the best strategy on their own, without communicating with their partner

node, the expression $\rho - \tau(q) > \frac{\rho - \tau(q') + \rho' - \tau(q)}{2}$ may be rearranged into $\rho - c(q) > \frac{\rho + \rho'}{2} - \frac{\tau(q) + \tau(q')}{2}$, from which the condition also holds.

Claim 4.2 and Lemmas 4.1 and 4.2 combined together complete the proof of the proposition (Fig. 4.5).

Remark Thus, the vehicular node interaction game satisfies Definition 4.2 for prisoner's dilemma kind of games.

When playing a prisoner's dilemma, each player must decide whether to cooperate with the opponent or whether to defect, assuming that each player does not know the decision of his opponent. The decision of each player is based on the following[1]: If a player believes that his opponent will cooperate, he should defect to maximize his payoff; in the vehicular node interaction game, this is the payoff labelled as A in Table 4.2. On the other hand, if a player believes that his opponent will defect, he should defect as well, because in such case cooperation will give less (cheating would give the payoff labelled C, whereas cooperation would give the payoff labelled D). This reasoning immediately implies:

Corollary 4.1 *In a single interaction, i.e. one-shot, of a prisoner's dilemma game, a best response strategy of both players is to defect.*

Corollary 4.2 follows:

Corollary 4.2 *In a single interaction of the vehicular node interaction game, a best response strategy of both players is to cheat.*

---

[1]Same reasoning as [4].

Corollaries 4.1 and 4.2 give rise to the question of how cooperation could be motivated, since the receiving and the sending nodes must cooperate in order to satisfy their corresponding needs for higher overall payoffs. Section 4.3 addresses this issue.

4.3   The Iterated Prisoner's Dilemma Model for the Repeated Interaction of Vehicular Nodes

Having shown that the vehicular node interaction model is in fact a prisoner's dilemma type of interaction, we propose a model for the iterative interaction model. The decision of whether to *cooperate* or *defect* in each interaction between any two vehicular nodes is modelled as an IPD game with *unknown* game horizon.

A repeated game makes it possible for the players to condition their moves on the complete previous history of the various stages, by employing strategies that define appropriate actions for each period. Such strategies are called *trigger strategies*. A trigger strategy is a strategy that changes in the presence of a predefined trigger; it dictates that a player must follow one strategy until a certain condition is met and then follow a different strategy, for the rest of the game.

One of the most popular trigger strategies is the *grim trigger strategy*, which dictates that the player participates in the relationship in a cooperative manner, but if *dissatisfied* for some known reason leaves the relationship forever. Exchanges based on such threats of non-renewing a relationship, which is based on a particular agreement between the two parties, are often referred to as *contingent renewal exchanges* [3]. Another popular strategy used to elicit cooperative performance from an opponent is for a player to mimic the actions of his opponent, giving the opponent the incentive to play cooperatively, since in this way he will be rewarded with a similar mirroring behaviour. This strategy is referred to as *tit-for-tat* strategy [2]. The choice for the cooperative iterative *grim* and *tit-for-tat* strategies results in an equilibrium (Fig. 4.6).

Fig. 4.6 In an iterated prisoner's dilemma, popular strategies include some type of reward or punishment in reaction to the opponent's play, e.g. grim strategy and tit-for-tat strategy

The situation we model is the following: mode j plays first and offers information to node $i \neq j$. Node i examines the information and makes a decision concerning whether to accept this as trusted information, according to i's employed iterative strategy. In the case that node i does not accept the information as trusted during the first interaction the game terminates with zero payoff to both i and j. If node i *trusts* the information, then the expected benefit from the interaction, U_i, is as defined in the single interaction game definition, i.e. $U_i = \pi(q) - \rho$. The evaluation of $\pi(q)$ is based on subjective measures and can be different for each vehicular node. Node i's decision to accept or trust information from another node induces the specific nodes to enter into a repetitive interaction of unknown horizon, having the options to cooperate with each other or to defect, in each interactive instance.

4.3.1 Subgame Perfect Equilibrium

In this situation, assume that node j plays first and offers information to node $i \neq j$. Node i examines the information and makes a decision concerning whether to accept this as trusted information according to i's employed iterative strategy. If node i does not accept the information as trusted during the first interaction, the game terminates with zero payoff to both i and j. If node i *trusts* the information, then the expected benefit from the interaction, U_i, is as defined in the single interaction game (i.e. $U_i = \pi(q) - \rho$). The evaluation of $\pi(q)$ is based on subjective measures and can be different for each vehicular node. Node i's decision to accept or trust information from another node induces the nodes to enter into a repetitive interaction of unknown horizon, having the options to cooperate with each other or to defect, in each interactive instance. Subsequently, we may interchange between referring to the player sending the information as *sending node* or j, and between referring to the player receiving the information as *receiving node* or i.

The analysis of the iterative vehicular node interactions employs the *grim* strategy as a cooperative strategy for the receiving node i and the *tit-for-tat* as a cooperative strategy for the sending node j. In addition, initially, one non-cooperative strategy is defined for each node. Specifically, the *cheat-and-leave* strategy is defined for i, while the *cheat-and-return* strategy is defined for j. The *cheat-and-leave* strategy is defined so as to allow the receiving node to employ non-cooperative behaviour (i.e. does not incur any resource sharing risk). With this strategy, the receiving node leaves the interaction after defecting from cooperation (i.e. does not continue interaction with the particular vehicular node to avoid any punishment for defecting from cooperation). The *cheat-and-return* strategy gives the opportunity to the sending node to defect from cooperation and not share information of high quality. Since it is the receiving node that accepts or rejects the interaction, the sending node can return to the interaction later; however, it must accept the receiving node's punishment, if any.

Consequently, the game profiles considered in the analysis involve combinations of these strategies. In the iterative vehicular node interaction game, it is possible to know the move of the opponent only after each interaction, since decisions during a game iteration are simultaneous. In order to compare different sequences of payoffs in iterative games, the idea of the present value (PV) of a payoff sequence is utilized. PV is the sum that a player is willing to accept currently instead of waiting for the future payoff (i.e. accept a smaller payoff today that will be worth more in the future) similar to making an investment in the current period that will be increased by a rate r in the next iteration. This is a popular method of evaluating a repetitive (possibly infinite or of unknown horizon) sequence of actions at a certain point in time.

Therefore, if a player's payoff in the next iteration were equal to 1, currently, the payoff the player would be willing to accept would be equal to $\frac{1}{1+r}$. Also, if there is a probability that the game (i.e. interaction) will not continue in the next iteration, equal to p, then the payoff the player is willing to accept currently (i.e. the player's PV) would be equal to $\delta = \frac{1-p}{1+r}$, where $\delta \in [0, 1]$ and often referred to as the *discount factor* in iterative games.

Therefore, given a payoff X in the next iteration, its PV in the current iteration equals $\delta \cdot X$. For an infinitely repeated game, a PV should include the discounted payoff of all subsequent periods of the game. Let the payoff from the current period be equal to 1. Then, the additional payoff a player is willing to accept for the next period equals to δ, for the period after the additional payoff equals to δ^2 and so on. Thus, PV equals to $1 + \delta + \delta^2 + \delta^3 + \delta^4 + \dots$, which, according to the sum of infinite geometric series, equals to $\frac{1}{1-\delta}$. Therefore, for a payoff X payable at the end of each period, the present value in an infinitely repeated game equals to $\frac{X}{1-\delta}$.

In addition to the sum of an infinite geometric series, we remind the reader of the method for finding the sum of a finite geometric series, as this is useful for the case when the game under study is expected to finish after a specific number of rounds, or, if we are interested in only studying a specific number of rounds. Therefore, the sum for a finite geometric series of n rounds equals to $\frac{X \cdot (1-\delta^n)}{1-\delta}$. Theorem 4.2 makes use of this, in addition to the sum for an infinite geometric series, which is also utilized for the proof of Theorem 4.1.

Since in repeated games there is an infinite number of decision nodes, they are described in terms of *histories* (i.e. records of all past actions that the players took [5]), thus a history corresponds to a path to a particular decision node in the infinitely repeated game tree. When a strategy instructs a player to play the best response to the opponent's strategy after every history (i.e. giving the player a higher payoff than any other action available after each particular history), it is called a *subgame perfect strategy*. When all players play their subgame perfect strategies, then we have an equilibrium in the repeated game, known as a *subgame perfect equilibrium*.

4.4 Defining a Repeated Vehicular Node Interaction Game

Definition 4.3 (Cheat-and-Leave Strategy) When the receiving node cooperates and then cheats in a random period, immediately leaving in the next period to avoid punishment, the strategy is referred to as the cheat-and-leave strategy.

Definition 4.4 (Cheat-and-Return Strategy) When the sending node cooperates and then cheats in a random period, immediately returning to cooperation in the next period, the strategy is referred to as the cheat-and-return strategy.

We are now ready to introduce a repeated game model of the vehicular node interactions when the history is taken into account in the decisions of the entities:

Definition 4.5 (Repeated Vehicular Node Interactions Game) Consider a game with infinite repetitions of the one-shot vehicular node interactions game with one additional action available to the receiving node: leaving the interaction.[2] Let the payoffs from each iteration be equal to the payoffs from the one-shot vehicular node interaction game, and in case the user leaves, let the payoff to both players be equal to zero. Then, the game is referred to as the *repeated vehicular node interaction game*.

Because of the unknown number of iterations, the PV of each player is calculated after a history (i.e. record of all past actions that the players made) to be able to evaluate each available action in the remaining game. As discussed earlier, the PV makes use of the idea of a discount factor $\delta = \frac{1-p}{1+r}$, where $p \in [0, 1]$ is the probability of termination of the interaction, and r is the rate of satisfaction gain of continuing cooperation in the next period (i.e. payoff increasing rate). Thus, the cumulative payoffs for each player from the repeated interaction can be considered.

Let the receiving node have a choice between the two following strategies: (1) the *grim* strategy (i.e. if cheating is perceived, then leave the relationship forever), and (2) the *cheat-and-leave* strategy. Let the sending node have a choice between: (a) the *tit-for-tat* strategy (i.e. mimic the actions of its opponent), and (b) the *cheat-and-return* strategy.

When neither of the two players cheats, the sequence of game profiles (from now on, refer to this simply as *profile*) is one of cooperation defined more formally next:

Definition 4.6 (Conditional-Cooperation Profile) When the receiving node employs the *grim* strategy and the sending node employs the *tit-for-tat* strategy, the profile of the repeated game is referred to as conditional-cooperation profile of the game.

---

[2]It is logical to assume that the user can switch to a different information source if dissatisfied, whereas a sending node does not want to leave the interaction once it has been selected as an information source.

The following theorem states that the *conditional-cooperation* profile strategies provide the best response to the alternative strategies: *cheat-and-leave* for the receiving node and *cheat-and-return* for the sending node.

Theorem 4.1 *In the repeated vehicular node interaction game, assume* $\delta > \frac{\tau(q)-\tau(q')}{\rho-\tau(q')}$ *and* $\delta > \frac{\rho-\rho'}{\pi(q)-\rho'}$ *then, the conditional-cooperation profile strategies result in higher payoffs than the* cheat-and-leave *and* cheat-and-return *strategies.*

Proof Assuming a history of cooperative moves in the past, the PVs of both the receiving and the sending nodes are computed. After comparing them, it is concluded that the conditional-cooperation profile strategies are more motivated than the *cheat-and-leave* and *cheat-and-return* strategies.

1. Assume that the receiving node i plays the grim strategy, the sending node's j could either play the tit-for-tat strategy (i.e. cooperate in the current period) or the cheat-and-return strategy, where it may cheat.

   Using the sum of infinite geometric series and the payoff at the end of each period as previously defined for the sending node, i.e. $\rho - \tau(q)$, then:
   If the sending node cooperates, then:

$$\mathrm{PV}_{coop}^{sn} = \frac{\rho - \tau(q)}{1 - \delta}.$$

   If the sending node defects, then:

$$\mathrm{PV}_{def}^{sn} = \rho - \tau(q') + \frac{\delta \cdot 0}{1 - \delta}.$$

   For the sending node to be motivated to cooperate, its PV in case of cooperation must be preferable than its PV in case of defecting. Thus:

$$\mathrm{PV}_{coop}^{sn} > \mathrm{PV}_{def}^{sn} = \frac{\rho - \tau(q)}{1 - \delta} > \rho - \tau(q') + \frac{\delta \cdot 0}{1 - \delta}.$$

   If the receiving node plays the grim strategy, the sending node is motivated to cooperate when $\delta > \frac{\tau(q)-\tau(q')}{\rho-\tau(q')}$.
2. Assume now that the receiving node i plays the cheat-and-leave strategy. Considering the sending node's j possible strategies, it could either cooperate or cheat in the current period.
   If the sending node cooperates, then:

$$\mathrm{PV}_{coop}^{sn} = \rho' - \tau(q) + \frac{\delta \cdot 0}{1 - \delta}.$$

   If the sending node cheats, then:

$$\mathrm{PV}_{def}^{sn} = \rho' - \tau(q') + \frac{\delta \cdot 0}{1 - \delta}.$$

If the receiving node plays the cheat-and-leave strategy, the sending node is not motivated to cooperate since $PV_{def}^{sn} > PV_{coop}^{sn}$

3. Assume now that the sending node j plays the tit-for-tat strategy. If the receiving node i plays the grim strategy, it will cooperate in the current period, whereas if it plays the cheat-and-leave strategy it may cheat.

Next, using the sum of infinite geometric series and the payoff at the end of each period as previously defined for the receiving node, i.e. $\pi(q) - \rho$, then:

If the receiving node cooperates, then:

$$PV_{coop}^{rn} = \frac{\pi(q) - \rho}{1 - \delta}.$$

If the receiving node cheats, then:

$$PV_{def}^{rn} = \pi(q) - \rho' + \frac{\delta \cdot 0}{1 - \delta}.$$

For the receiving node to be motivated to cooperate, its PV in case of cooperation must be preferable than its PV in case of cheating. Thus:

$$PV_{coop}^{rn} > PV_{def}^{rn} = \frac{\pi(q) - \rho}{1 - \delta} > \pi(q) - \rho' + \frac{\delta \cdot 0}{1 - \delta}.$$

If the sending node cooperates, the receiving node is motivated to cooperate when $\delta > \frac{\rho - \rho'}{\pi(q) - \rho'}$.

4. Finally, assume that the sending node j plays the cheat-and-return strategy. Considering the receiving node's i possible strategies, it could either cooperate or cheat in the current period.

If the receiving node cooperates, then:

$$PV_{coop}^{rn} = \pi(q') - \rho + \frac{\delta \cdot 0}{1 - \delta}.$$

If the receiving node defects, then:

$$PV_{def}^{rn} = \pi(q') - \rho' + \frac{\delta \cdot 0}{1 - \delta}.$$

If the sending node plays the cheat-and-return strategy, the receiving node is not motivated to cooperate since $PV_{def}^{rn} > PV_{coop}^{rn}$.

From above, it follows that the conditional-cooperation profile is motivated when $\delta > \frac{\tau(q) - \tau(q')}{\rho - \tau(q')}$ and $\delta > \frac{\rho - \rho'}{\pi(q) - \rho'}$.

4.4.1   Using an Adaptive Strategy to Motivate Cooperative Behaviour

Within the context of the vehicular node interaction game, the interaction model payoffs should reflect a node's preference *history*, where continuous cooperative or trusted behaviour may be reflected as a lower risk value for a specific node, whereas continuous defective behaviour can be represented as a higher risk value. We employ the idea of an adaptive player, such that a node i 's decision of whether to trust node $j \neq i$, or not, considers a j 's probability to stay trustworthy based on the acquired knowledge gained by i over the course of the repeated game.

Thus, when a node must evaluate the expected benefit from a specific interaction, the evaluation should consider the probability of remaining trustworthy, calculated dynamically from observing past node behaviour. Being an adaptive player, the node makes a more informed decision on whether to cooperate or defect. This is achieved by multiplying the expected benefit $\pi(q)$ by a variable α (Definition 4.7), representing the estimated probability of trustworthy behaviour. To achieve this, we assume that any node i has an internal state, which modifies probability α after every interactive period with node $j \neq i$, and that α has a different value for each different network that interacts with the user.

Definition 4.7   In the vehicular node interaction game, node i possesses an internal state, which, based on a history of behaviour from node $j \neq i$, estimates node j's probability not to demonstrate degradation, $\alpha \in [0, 1]$. Given that node i has an expected behaviour e, from node j, such that $\pi(q) \geq e$, the value of α at the end of an interactive period is modified according to (4.1).

$$\alpha_{now} = \begin{cases} \alpha_{previous} + (\alpha_{previous} \cdot \frac{\pi(q)_{final} - e}{\pi(q)_{final}}), \\ \quad \text{if } \pi(q)_{final} \geq e \; \& \; \alpha_{now} \leq 1 \\ 1, \\ \quad \text{if } \pi(q)_{final} \geq e \; \& \; \alpha_{now} \geq 1 \\ \alpha_{previous} \cdot \frac{\pi(q)_{final}}{e}, \\ \quad \text{otherwise} . \end{cases} \tag{4.1}$$

By introducing the variable $\alpha \in [0, 1]$, node i considers node j's history, approaching the decision of whether to cooperate or defect in an adaptive manner, i.e. by evaluating $\pi(q) \cdot \alpha$ instead of only $\pi(q)$.

In Sect. 4.3 the selection of the *grim* and *tit-for-tat* strategies is proposed for a cooperative game profile, i.e. node i punishes node j forever if deceptive behaviour is perceived even once, or, node i punishes node j with only one period of absence, even if node j demonstrates degradation frequently. Considering the adaptive way in which node i takes a decision with the use of α, we propose a new strategy to be employed as a means to interact with the selected network.

Let the strategy be the following: cooperate as long as the interacting node cooperates; if the interacting node cheats, then leave for a x number of periods; after that, return and cooperate again. Let the number x be equal to 1 if $\alpha = 1$ or $\frac{1}{\alpha}$ otherwise; such that a node with a lower value for α suffers a separation of more periods with the user, whereas a network with a higher value for α is punished for less periods (minimum punishment is 1 period).

Definition 4.8 The adaptive-return strategy dictates that if a node perceives deception, then the opponent is punished by the node leaving for a x number of periods, before returning back to cooperation. The value of x is a user-generated value as defined in (4.2).

$$
x = \begin{cases} 1, & \text{if } \alpha = 1 \\ \lceil \frac{1}{\alpha} \rceil, & \text{otherwise .} \end{cases} \tag{4.2}
$$

Next, we show that when a node employs the *adaptive-return* strategy, the interacting node is at least as motivated to cooperate as when the one-period punishment strategy is employed.

Theorem 4.2 *Given a history of cooperation between any two vehicular nodes, the conditions necessary to impose cooperation when a node employs the adaptive-return strategy are also necessary to impose cooperation when the profile of the game is the one-period punishment profile.*

Proof Given a history of the game where both players have cooperated in the past, a node has two options in the current period: cooperate and cheat. When the node cooperates, the present value is as follows:

$$
PV_{cooperate} = \frac{(\rho - \tau(q)) \cdot 1 - \delta^{x+1}}{1 - \delta} + \frac{\delta^{x+2} \cdot \rho - c\tau(q)}{1 - \delta}.
$$

The sum of a finite geometric progression is used to calculate the discounted value for the first $x + 1$ periods. If the node cheats, the present value is

$$
PV_{cheat} = \rho - \tau(q') + \frac{1 - \delta^x \cdot 0}{1 - \delta} + \frac{\delta^{x+2} \cdot \rho - \tau(q)}{1 - \delta}.
$$

The conditions necessary to impose cooperation are calculated next:

$$
PV_{cooperate} > PV_{cheat} = \frac{(\rho - \tau(q)) \cdot 1 - \delta^{x+1}}{1 - \delta} + \frac{\delta^{x+2} \cdot \rho - \tau(q)}{1 - \delta}
$$

$$
> \rho - \tau(q') + \frac{1 - \delta^x \cdot 0}{1 - \delta} + \frac{\delta^{x+2} \cdot \rho - \tau(q)}{1 - \delta}.
$$

Now consider the case that the node cooperates in the current period. Then, the interacting node has two options: to cooperate or to defect from cooperation.

If the interacting node cooperates, then:

$$\text{PV}_{coop}^{net} = \rho - \tau(q) + \delta \cdot (\rho - \tau(q)) + \frac{\delta^2 \cdot (\rho - \tau(q))}{1 - \delta}.$$

If the interacting node defects, then:

$$\text{PV}_{def}^{net} = \rho - \tau(q') + \delta \cdot 0 + \frac{\delta^2 \cdot (\rho - \tau(q))}{1 - \delta}.$$

In order for cooperation to be motivated, it must be that:

$$\text{PV}_{coop} > \text{PV}_{def} =$$

$$\rho - \tau(q) + \delta \cdot (\rho - \tau(q)) + \frac{\delta^2 \cdot (\rho - \tau(q))}{1 - \delta} >$$

$$\rho - \tau(q') + \delta \cdot 0 + \frac{\delta^2 \cdot (\rho - \tau(q))}{1 - \delta}.$$

Simplifying, we get $\delta > \frac{\tau(q) - \tau(q')}{\rho - \tau(q)}$.

Since $\delta \in (0, 1)$, $x \geq 1$, and $\delta^{x+1} \leq \delta^2$, such that $\delta_{x+1} - 1 \leq \delta^2 - 1$, we may simplify for δ to get $\delta > \frac{\tau(q) - \tau(q')}{\rho - \tau(q)}$, similarly to the value of δ when employing the one-period punishment strategy. In fact the simplification of $\delta^{x+1} \leq \delta^2$ considers the case that the adaptive-return strategy generates a punishment of only one period, whereas if we consider greater values of x, the motivation to cooperate increases, showing that the conditions for sustaining cooperation when the adaptive-return strategy is used by the user are necessary for sustaining cooperation under the one-period punishment profile.

4.5   Evaluating User–Network Cooperation

This section examines the numerical behaviour of the sending and receiving node strategies defined and used for the repeated game defined. The evaluation is based on a Matlab implementation of the repeated game, where all sending and receiving node strategies are played against each other multiple times in order to evaluate the behaviour of each strategy in terms of payoffs. The implementation of the game model was based on a publicly available Matlab implementation of the iterated prisoner's dilemma Game [6], which has been extended to include all existing and proposed strategies examined in this chapter.

The implementation makes use of the following guidelines, set to reflect the analytical model of the repeated game. In each simulation run, both players play

Table 4.3 Payoffs for the different simulation sets

Strategies	Simulation set 1	Simulation set 2	Simulation set 3
Player leaves	(0, 0)	(0, 0)	(0, 0)
One defects, one cooperates	(4, 1)	(100, 1)	(90, 10)
Both defect	(2, 2)	(40, 40)	(50, 50)
Both cooperate	(3, 3)	(60, 60)	(60, 60)

their strategies and get payoffs accordingly. In the first set of simulations the payoffs are the following: when a player leaves, they both get 0 in the specific period, if one defects and one cooperates, the first gets 4 and the other gets 1, if they both defect, each gets 2, and if they both cooperate each gets 3. In the second set of simulations, we investigate the behaviour of the players when the difference between defecting and cooperating increases. The payoffs for the second set of simulations are the following: when a player leaves, they both get 0 in the specific period, if one defects and one cooperates, the first gets 100 and the other gets 1, if they both defect, each gets 40, and if they both cooperate each gets 60. In the third set of simulations, we investigate the behaviour of the players when there is a small difference in the payoffs between the cases of both defecting versus both cooperating. The payoffs for the third set of simulations are the following: when a player leaves, they both get 0 in the specific period, if one defects and one cooperates, the first gets 90 and the other gets 10, if they both defect, each gets 50, and if they both cooperate each gets 60.

We use simple numbers as payoffs to help us get some scores for different strategy combinations but these numbers follow the relationships of the payoffs as described in their general case in the repeated game model (Table 4.2). The payoffs for the different simulation sets are summarized in Table 4.3.

Furthermore, for the adaptive strategy, the value of α is randomly generated at the beginning of each simulation run but adapted according to the players' behaviour during the actual simulation (since each run simulates an iterative process). A randomly generated satisfaction and a fixed threshold of expected satisfaction are also implemented for the strategies in all simulation runs.

A randomly generated number of iterations was run for each set of simulations to get cumulative sending and receiving payoffs for each combination of strategies. The player payoffs per strategy are eventually added to give the most profitable sending and receiving network strategies, respectively, for the total number of iterations of a simulation run; then, the average cumulative payoffs from all simulation runs are calculated. Although the number of iterations is randomly generated, we still repeat the process 100 times for each set of simulations, i.e. by randomly generating 100 different numbers of iterations, in order to include behaviours when the number of iterations is both small and large.

Table 4.4 Receiving node payoffs from all strategy combinations (1st simulation set)

	Rec. node payoffs	
	Sending node strategies	
Rec. node strategies	Tit-for-tat	Cheat&return
Grim	793.14	4.42
Cheat&leave	6.62	4.65
Leave&return	793.14	252.92
Adaptive-return	793.14	355.05

Table 4.5 Sending payoffs from all strategy combinations (1st simulation set)

	Send. node payoffs	
	Network strategies	
Rec. node strategies	Tit-for-tat	Cheat&return
Grim	793.14	7.42
Cheat&leave	3.82	4.23
Leave&return	793.14	617.21
Adaptive-return	793.14	618.57

For the first set of simulations, i.e. with the payoffs ranging from 0 to 4, the payoffs are calculated for an average of 264.38 iterations per simulation run.[3] In both tables we see a score for each strategy combination. The score corresponds to either a receiving node payoff (Table 4.4) or a sending node payoff (Table 4.5).

The results for the first set of simulations show that the most profitable receiving node strategy is the *adaptive return* strategy, and that the most profitable sending node strategy is the *tit-for-tat* strategy for all payoffs except for the payoff received from the combination with the *cheat&leave* strategy of the receiving node. However, the difference between the payoffs received by the sending node from playing either the *tit-for-tat* strategy or the *cheat&return* strategy, in combination with the *cheat&leave* strategy of the receiving node, is negligible. Furthermore, the combination of the two most profitable strategies in the same game profile gives the highest cumulative payoffs to both players. Based on these numerical results, we define the *adaptive-punishment* profile (Definition 4.9) for the game to consist of the *adaptive-return* strategy for the user and the *tit-for-tat* strategy for the network.

Definition 4.9 (Adaptive-Punishment Profile) When the receiving node employs the *adaptive-return* strategy and the sending node employs the *tit-for-tat* strategy, the profile of the repeated game is referred to as *adaptive-punishment* profile of the game.

For the second set of simulations, i.e. with the payoffs ranging from 0 to 100, the payoffs are calculated for an average of 240.59 iterations per simulation run.[4] As

---

[3]Minimum iterations generated: 8, maximum iterations generated: 1259.

[4]Minimum iterations generated: 2, maximum iterations generated: 1124.

Table 4.6 Receiving node payoffs from all strategy combinations (2nd simulation set)

	Rec. node payoffs	
	Sending node strategies	
Rec. node strategies	Tit-for-tat	Cheat&return
Grim	14,435.4	47.8
Cheat&leave	164.2	108.36
Leave&return	14,435.4	4894.76
Adaptive-return	14,435.4	4974.47

Table 4.7 Sending node payoffs from all strategy combinations (2nd simulation set)

	Sending node payoffs	
	Sending node strategies	
Rec. node strategies	Tit-for-tat	Cheat&return
Grim	14,435.4	146.8
Cheat&leave	65.2	95.49
Leave&return	14,337.4	12,850.4
Adaptive-return	14,435.4	12,861.8

Table 4.8 Receiving node payoffs from all strategy combinations (3rd simulation set)

	Rec. node payoffs	
	Sending node strategies	
Rec. node strategies	Tit-for-tat	Cheat&return
Grim	16,416.6	62.7
Cheat&leave	159.9	100.7
Leave&return	16,416.6	6423.6
Adaptive-return	16,416.6	6449.7

previously, we see in both tables a score for each strategy combination. The score corresponds to either a receiving node payoff (Table 4.6) or a sending node payoff (Table 4.7).

Again, we observe that for the second set of simulations, the most profitable strategy for the receiving node is the *adaptive-return* strategy and the sending node's most profitable strategy is the *tit-for-tat* strategy.[5] The increase in the differences between cooperating and defecting payoffs resulted in higher overall payoffs but has not changed the general payoff trend for the two players. In total, the highest payoffs are experienced by the players when they decide to use cooperating strategies instead of defecting; this result motivates the players to go ahead and cooperate.

For the third set of simulations, i.e. with the payoffs ranging from 0 to 90, the payoffs are calculated for an average of 268.6 iterations per simulation run.[6] As previously, we see in both tables a score for each strategy combination. The score corresponds to either a receiving node payoff (Table 4.8) or a sending node payoff (Table 4.9).

---

[5]Except for the payoff in combination with the receiving node's *cheat&leave* strategy, however, the preferred profile for the game is still the *adaptive-punishment* profile.

[6]Minimum iterations generated: 2, maximum iterations generated: 1342.

Table 4.9 Sending node payoffs from all strategy combinations (3rd simulation set)

	Sending node payoffs	
	Sending node strategies	
Rec. node strategies	Tit-for-tat	Cheat&return
Grim	16,416.6	141.9
Cheat&leave	80.7	76.7
Leave&return	16,416.6	13,709.7
Adaptive-return	16,416.6	13,710

Again, for the third set of simulations, the most profitable strategy for the receiving node is the *adaptive-return* strategy and the sending node's most profitable strategy is the *tit-for-tat* strategy. Overall, the preferred profile for the game is still the *adaptive-punishment* profile. The small difference between the cases of both players defecting and of both players cooperating has not changed the general payoff trend for the two players. In total, the highest payoffs are experienced by the players when they decide to use cooperating strategies instead of defecting; this result continues to motivate the players towards cooperation.

Furthermore, it is worth noting that, for all simulation sets, when the receiving node plays the *leave&return* strategy, the payoffs received by the receiving node are comparable (though less) to the highest payoffs received, i.e. payoffs for employing the *adaptive-return* strategy. The justification for these results is the following: we have shown that the minimum conditions for the sending node to cooperate when the receiving node employs the *adaptive-return* strategy are also necessary for the sending node to cooperate when the *leave&return* strategy is employed. In fact, the conditions for the two strategies are at least equal, and given this, it is expected to observe payoff values that are numerically closer compared to payoff values for the other receiving node strategies, although it appears that the *adaptive-return* strategy manages to achieve slightly higher payoffs than the *leave&return* strategy.

An additional observation is that the *leave&return* strategy is weaker in the case the sending node decides to defect, because the receiving node's reaction is a fixed-period punishment. On the other hand, when the receiving node employs the *adaptive-return* strategy, the punishment period is not fixed but adapts to the sending node's past behaviour, appearing to consequently achieve a slight improvement in the overall receiving node's payoff.

An additional factor that we should consider when interpreting the obtained payoffs is that each simulation investigates a single game interaction, thus not considering dependencies that may arise in an evaluation of multiple co-existing interactions, where a receiving node may interact with several sending nodes and vice versa.

4.6 Conclusion

In general, trusting a perfect stranger is a complicated decision-making process, let alone while driving. The highly dynamic nature of vehicular networks and the absence of a trusted third party make it even harder to take such a decision. Instituting the trustworthiness of this type of ad hoc interaction is essential to successfully establish a social relationship, especially in case that adaptive-type of strategies are used to motivate cooperation. In this chapter, a game theoretic approach has been defined to mathematically motivate trusted and cooperative interactions among users in vehicular networks by harnessing collective intelligence, especially through the use of feedback from the vehicular network participating nodes, in an adaptive manner.

In fact, within the context of the vehicular node interaction game, especially the repeated flavour of the game elaborated above, the interactions model payoffs should reflect the node's preference *history*, where continuous cooperative or trusted behaviour may be reflected as a lower risk value for a specific node, whereas continuous defective behaviour can be represented as a higher risk value. The chapter adopts this approach as a new type of player in the game, the adaptive player. A player i undertaking this adaptive profile acts such that a node i's decision of whether to trust node $j \neq i$, or not, considers the probability of j's staying trustworthy based on the acquired knowledge gained by i over the course of the repeated game (Fig. 4.7).

Fig. 4.7 The adaptive strategy has been shown to be very effective in the vehicular node interaction game, as it allows the players to adjust their actions according to their opponent's level of cooperative behaviour

The significance of this particular profile in the resolution of the game comes from the fact that when a node must evaluate the expected benefit from a specific interaction, the evaluation should consider the probability of remaining trustworthy, calculated dynamically from observing past node behaviour.

References

1. Axelrod, R.M.: The Evolution of Cooperation. Basic Books, New York (1984)
2. Dixit, A., Skeath, S.: Games of Strategy. W.W. Norton & Company, New York (1999)
3. Gintis, H.: Game Theory Evolving: A Problem-Centered Introduction to Modeling Strategic Interaction. Princeton University Press, Princeton (2000)
4. Kendall, G., Yao, X.: The Iterated Prisoner's Dilemma: 20 Years On. Advances in Natural Computation Book Series, vol. 4. World Scientific Publishing, Singapore (2009)
5. Myerson, R.B.: Game Theory: Analysis of Conflict. Harvard University Press, Cambridge (2004)
6. Taylor, G.: Iterated prisoner's dilemma in MATLAB. Archive for the "Game Theory" Category (2007). http://maths.straylight.co.uk/archives/category/game-theory

Chapter 5
Using Game Theory to Characterize Trade-Offs Between Cloud Providers and Service Providers for Health Monitoring Services

Abstract In this chapter, we consider a 5G network, where service providers offer health monitoring services, by making use of a wearable devices, which are placed on the customers such that they are able to collect personal data for each customer, by monitoring selected health indicators, e.g. the heart rate. A consideration for this scenario is that the continuous generation of new data makes the complete set of data that needs to be manipulated, such a large amount, that it is impossible to store it locally, but alternatively, it becomes necessary to store this data away from the user and make it accessible through the Internet. This is usually done by using storage on the cloud, which the health-related data can use as storage. The use of the cloud offers the additional advantage for the service provider of the health monitoring service that the data can be easily accessible through the use of the cloud over the Internet. Therefore, we must consider the interaction between the health monitoring service providers and the cloud providers over the handling of the user's data. In particular, we consider the case where enhanced data protection demands by the user for the particular health service may require the cooperation of the service provider and the cloud provider in advance. Once the agreement is reached that the service provider and the cloud provider prefer to cooperate rather than not for the particular monitoring service, a bargaining scheme is employed to allow them to reach an agreement with regard to their individual revenues.

Keywords Game theory · Cloud provider · Service provider · Health monitoring service · Bargaining model

5.1 Overview of Technology and of Relevant Game-Theoretic Tools

In a 5G network, service providers offer health monitoring services (often based on artificial intelligence algorithms, i.e. AI-based health monitoring services), by making use of a special type of devices that the customers *wear* continuously, also known as wearable devices. Wearable devices are fashioned to collect a large

© Springer Nature Switzerland AG 2020
J. Antoniou, *Game Theory, the Internet of Things and 5G Networks*,
EAI/Springer Innovations in Communication and Computing,
https://doi.org/10.1007/978-3-030-16844-5_5

Fig. 5.1 Service providers offer health monitoring services often based on artificial intelligence algorithms

amount of data for each customer, by monitoring selected health indicators, e.g. the heart rate. The continuous generation of new data makes the complete set of data that needs to be manipulated, such a large amount, that it is impossible to store it locally, either on the monitoring device or on a personal device owned by the user, e.g. a mobile phone. It becomes necessary to store this data away from the user and make it accessible through the Internet. This is usually done by using storage on the cloud, a global network of servers that interoperate to form a single ecosystem, which is used, among other functions, to store data. Therefore, the health-related data can use the cloud ecosystem as storage, which also is an advantage for the service provider of the health monitoring service, because the data can be easily accessible through the use of the cloud (Fig. 5.1).

The cloud solution for storing the health-related data does not come without risks since the cloud is administered by a cloud provider and not the service provider, who has the original agreement with the user. Thus, essentially two entities have now access to the user personal and sensitive health data, the health monitoring service provider (HMSP) and the cloud provider (CP). Note that agreements between the service and cloud providers are often made to prevent unethical access to, and subsequently unethical use of data. Furthermore, the data that we are dealing with in this sort of applications is not only personal data but often sensitive data and must be handled with care, as many legislations exist that protect from misuse of the user's data (e.g. consent forms must be made available for the user to agree to any personal data manipulation, etc.). It seems that for this scenario to happen, the HMSP provider and the CP must come to some sort of an agreement. This type of negotiation and agreement between the HMSP and the CP will be further explored in the following sections that deal specifically with the details of this scenario (Fig. 5.2).

5.1.1   Overview of Bargaining and Bayesian Games

With the exception of the groundbreaking contributions of John F. Nash [4], illustrated in Fig. 5.3, bargaining theory is mainly based on [3], which basically evolved from the seminal paper by Ariel Rubinstein [6], who succeeded in making the procedure of bargaining quite attractive, mainly due to the proposed model's simplicity and ease of understanding.

Fig. 5.2 Big data, i.e. enormous amounts of data, often use storage on the cloud, a global network of servers that interoperate to form a single ecosystem, which is used, to store data, on behalf of end devices with less storage capacity

Fig. 5.3 Graphical representation of the groundbreaking contribution of John F. Nash in bargaining theory from his original paper entitled *The Bargaining Problem* [4]

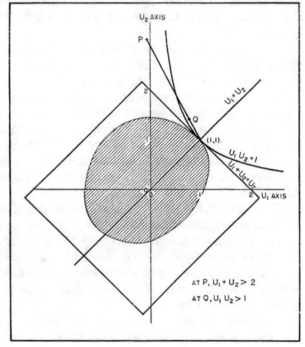

Figure 5.3, shows the anticipated utilities of two bargaining individuals, at the agreements point of the bargaining problem. While the author uses the graph to demonstrate the agreement or equilibrium point, the paper further illustrates a set of examples to showcase the application of the theory, which are reproduced in Table 5.1. Specifically, the example deals with a set of goods over which two individuals attempt to trade, negotiating over the goods while knowing the utility

Table 5.1 Illustration of an example of a bargaining situation as presented in [4]

	Utility to bill	Utility to jack
Bill's goods		
Ball	2	4
Whip	2	2
Ball	2	1
Bat	2	2
Box	4	1
Jack's goods		
Pen	10	1
Toy	4	1
Knife	6	2
Hat	2	2

of each good to each player. The paper [4] refers to the two individuals as *Bill* and *Jack*, and the result of the negotiation (proving also the solution point in Fig. 5.3) is that *Bill gives Jack: book, whip, ball, and, bat* and that *Jack gives Bill: pen, toy, and, knife* (see 5.1).

More specifically, a bargaining situation is an exchange situation, in which two individuals have a common interest to *negotiate* a product or a service but simultaneously have conflicting interests about the *price point* at which to trade, because the seller, offering the service, would like to trade at a higher price, while the buyer, aiming to acquire the service, would like to trade at a lower price. Therefore, in a bargaining situation, the two players have a common interest to cooperate but have conflicting interests about exactly how to cooperate. On the one hand, each player would like to reach an agreement rather than disagree; on the other hand, each player wants to reach an agreement that is as satisfying to that player's request as possible.

Therefore, a bargaining situation may be modelled as a game situation since the outcome of bargaining depends on both players' bargaining strategies, i.e. whether or not an agreement is reached and the terms of that agreement (if one is reached) depend on both players' actions during the bargaining process. There exist two well-known game models of a bargaining situation between two players, which will describe briefly next, the Nash bargaining game model and the Rubinstein bargaining game model.

The Nash bargaining game model defines a solution (known as the Nash bargaining solution) by a simple formula and it is applicable to a large class of bargaining situations. The classic Nash bargaining game example is the scenario where two players bargain over the partition of a cake of fixed size. Since the cake will be partitioned to the two players, the addition of their partitions should equal the total cake; therefore, the set of non-zero partitions, which sum to the total amount of the cake, is the set of possible agreements in the bargaining situation. In the case of disagreement, each player receives a penalty, which is defined according to the bargaining situation under consideration; the definition of a penalty comes from the fact that in bargaining situations the desirable outcome is agreement, thus disagreement results in a non-satisfactory payoff for the two players. The penalties of the two players are defined as the *disagreement point* of the game.

The *useful payoff* for each player may be defined to be the player's payoff received from the received partition in case of agreement, minus the penalty that would be received in case of disagreement as defined in the *disagreement point*. The unique solution of the cake partition game is therefore, the unique pair of partitions that maximizes the product of the players' *useful payoffs* and is referred to as the *Nash product* or the *Nash bargaining solution*.

The second game scenario that we will describe is known as the Rubinstein bargaining game. This is a game model that is modelled as a sequential-moves game, in which the players take turns to make offers to each other until agreement is secured. This model is an intuitive model, because a lot of real-life negotiations are based on the idea of making offers and counter-offers. From the sequential-moves model of the bargaining process, it is easy to see that if the two negotiating players do not incur any costs or penalties for delaying the agreement decision, the solution to the bargaining game may be indeterminate, because the two players could continue to negotiate forever. Given that there is a cost to each player for delaying, then each player's bargaining power is determined by the magnitude of this cost.

With regard to the Rubinstein game model that if we consider that the negotiations take place over the partition of a cake, similarly to the Nash bargaining game model, then we can compare the two game solutions. Considering the moves of the players in a sequential manner, we have the following: the first player proposes a partition of the cake; if the second player accepts, agreement is reached and the game is over; otherwise, the second player proposes a different partition, and the process of alternating offers continues until an offer is accepted. However, for each additional negotiation round there is a cost to each player, a cost could be that the size of the cake becomes smaller in each round. The factor by which the cake gets smaller may be different for each player and it is referred to as the player's discount factor. The Rubinstein bargaining game model has a unique subgame perfect equilibrium, which makes use of the fact that any offer made at each round by a player should be equal or greater to the discounted best value that the opponent can get in the next round.

Another element that must be considered in bargaining situations is truthfulness of the players. In order to motivate the two bargaining players to be truthful about any information that may affect the bargaining process, there must exist a mechanism that can penalize a player who turns out to be dishonest about the real cost or discount factor, assuming that it is detectable whether a player has been dishonest or not. Such mechanisms exist and are called *pricing mechanisms* [5], constituting an interesting and very promising way to guarantee the truthfulness of the participating players. In relevant literature there is a particular research field that is focused on the development, limitations and capabilities of such pricing mechanisms, known as the algorithmic mechanism design [5]. The algorithmic tools and theoretical knowledge that have already developed in the field of algorithmic

Fig. 5.4 In consideration of various bargaining situations and model solutions, an element that must be considered and subsequently verified is the truthfulness of the players

mechanism design constitute a fruitful pool for extracting algorithmic tools for enforcing players to act honestly, through pricing mechanisms, once these tools are customized and further developed for handling the needs of any specific scenario, e.g. a bargaining scenario in this case (Fig. 5.4).

In several situations where interactions occur, the interacting players may not have complete information about each other's characteristics. The model of a Bayesian game is designed to model such situations. A player's uncertainty about the opponent's properties is modelled by introducing a set of possible states, i.e. probable sets of characteristics that a player may have, also known as a player's *types*. Each player assigns a probability of occurrence to each of the opponent's possible *types*. Consequently, a definition of a Bayesian game is similar to the definition of a normal interaction game between a number of players, with the additional consideration of players' *types* and the corresponding probability of occurrence, as believed by the opponent(s).

A Bayesian game can be modelled by introducing nature as a player in the game. This approach was proposed by John C. Harsanyi in [2]. Nature assigns a random variable to each player, which could take values of types for each player and associate probabilities with these types. In the course of the game, nature randomly chooses a type for each player according to the probability distribution across each player's types. The type of a player determines the player's payoff and the fact that a Bayesian game is a game model of incomplete information means that at least one player is unsure of the type and thus the payoff of another player.

In any given play of a Bayesian game, players know their type and in determining their best action, they must consider what the other player(s) would do if any of the other possible types were to occur, since any player may be imperfectly informed about the current state of the game. Therefore, a Nash equilibrium of a Bayesian game is the Nash equilibrium of the normal form game in which the set of players includes all possible types for each player, and consequently the set of actions includes all possible actions for each such state of every player considered. For a

Nash equilibrium to be determined in a Bayesian game, each player must choose the best strategy, given his belief about the occurrence of the opponent's types, the state of the game and the possible strategies of the opponent players.

5.1.2 Introducing the Bargaining Scenario Between the HMSP and the CP

As previously mentioned, we consider a 5G network, where service providers offer health monitoring services, by making use of wearable devices, which are placed on the customers such that they are able to collect personal data for each customer, by monitoring selected health indicators, e.g. the heart rate. A consideration for this scenario is that the continuous generation of new data makes the complete set of data that needs to be manipulated, such a large amount, that it is impossible to store it locally, but alternatively, it becomes necessary to store this data away from the user and make it accessible through the Internet. This is usually done by using storage on the cloud, which the health-related data can use as storage. The use of the cloud offers the additional advantage for the service provider of the health monitoring service that the data can be easily accessible through the use of the cloud over the Internet. Therefore, we must consider the interaction between the health monitoring service providers (HMSP) and the cloud providers (CP) over the handling of the user's data (Fig. 5.5).

In this chapter, we consider the case where enhanced data protection demands by the users for the particular health service may require the cooperation of the

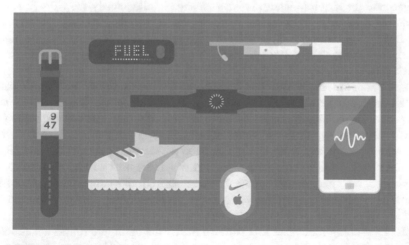

Fig. 5.5 Health monitoring services make use of wearable devices, which are placed on the users such that they are able to collect personal data for each customer, by monitoring selected health indicators, e.g. the heart rate

HMSP and the CP in advance to support a particular user. The HMSP and the SP are selected such that the HMSP has a contract with the user and the CP has a contract with the CP with regard to the user's data storage and manipulation requirements. Once the selection is completed, the HMSP and the CP are required to cooperate regarding their revenues. The user makes a payment to the HMSP and in turn the HMSP makes a payment to the CP, and thus the user's payment is indirectly partitioned between the HMSP and the CP. The payment partition is basically a pre-calculated configuration that is adopted by the HMSP and the CP, with reference to the particular health monitoring service, to ensure that the whole process is transparent to the user, obeying the user-centric data protection requirements dictated for such services in 5G networks.

5.2 The Health Monitoring Service Scenario Elaborated

Consider the case where a customer of the converged mobile communication networks makes a service request for a particular health monitoring service with strict data protection requirements especially where it comes to data storage and manipulation. Both the HMSP and the CP will provide resources to satisfy this service request. Further consider that it is advantageous for two providers, i.e. the HMSP and the CP to cooperate in order to offer the user a higher guarantee of service delivery with the particular data protection requirements satisfied. The cooperation entails that in order to support the particular user's health monitoring service request, the HMSP will serve the request in terms of manipulating the data to generate the required health indicators, and simultaneously the CP will reserve resources necessary to efficiently store the data generated, making it available to the HMSP. This cooperation enables service provision to be offered at the highest guarantees in terms of service to the user and in terms of data protection for the user's data.

Since both the HMSP and the CP may quantify their satisfaction of participating in the service in terms of their revenue gain, and since the two providers must cooperate for the single service to the user, then the payment for supporting the service needs to be partitioned between them, in order for the networks to have an incentive to cooperate. Note that this may be pre-decided between the HMSP and the CP, which could in turn set the price at which the user will be charged for the service, but our analysis will approach this scenario from a cooperation perspective. Therefore, from the cooperation perspective, the partitioning configuration must be such that it is satisfying to both providers.

This reasoning motivated this work to model the configuration as a solution to a cooperative bargaining game, which is presented next. The scenario defines the *payment partition* as a game of bargaining between the two providers, specifically the HMSP and the CP. Firstly, we define the payment partition as a game and, then, we show that this is equivalent to the well-known Rubinstein bargaining game [3], when the agreement is reached in the first negotiation period. Given this equivalence, an optimal solution to the Rubinstein bargaining game would also constitute an optimal solution to the payment-partition game. The resolution of the game is presented in Sect. 5.3.

Let $q \in Q$ be the data protection guarantees level for which the two providers negotiate. Consider the payment-partition scenario, where two providers want to partition a service payment $\pi_i(q)$ paid by the health monitoring service customer. Let $c_i(q)$ be considering the cost of provider i. Note that this is known to each provider, e.g. this could represent a resource reservation cost for the CP, whereas it could represent a computational cost for the HMSP.

Given the cost characteristics of provider i, each provider seeks a portion:

$$\pi_i(q) = c_i(q) + \phi_i(q), \tag{5.1}$$

where $\phi_i(q)$ is the actual profit of provider i, such that:

$$\pi_1(q) + \pi_2(q) = \Pi(q), \tag{5.2}$$

where $\Pi(q)$ is the total payment paid by the health monitoring service customer. The providers' goal is to find the payment partition, which will maximize the value of $\phi_i(q)$, given the values of $\Pi(q)$ and $c_i(q)$. Definition 5.1 defines the bargaining game between the two providers:

Definition 5.1 (Payment-Partition Game) Fix a specific data protection guarantees level $q \in Q$ such that a fixed payment Π is received. Consider a one-shot strategic game with two players corresponding to the two providers. The profiles of the game, i.e. the strategy sets of the two players, are all possible pairs (π_1, π_2), where $\pi_1, \pi_2 \in [0, \Pi]$ such that $\pi_1 + \pi_2 = \Pi$. All such pairs are called agreement profiles and define set S^a. So, $S^a = \pi_1 \times \pi_2$. In addition, there exists a so-called disagreement pair $\{s_1^d, s_2^d\}$, which corresponds to the case where the two players do not reach an agreement. So, the strategy set of the game is given by $\mathscr{S} = S^a \bigcup \{s_1^d, s_2^d\}$. For any agreement point $s \in S^a$ the payoff $U_i(s)$, for player $i \in [2]$, is defined as follows:

$$U_i(s) = \pi_i - c_i. \tag{5.3}$$

Otherwise,

$$U_1(s_1^d) = U_2(s_2^d) = 0. \tag{5.4}$$

This game is referred to as the payment-partition game.

Fact

Let $s^* = (\pi_1^*, \pi_2^*)$ be an optimal solution of the payment-partition game. Then $\phi_i = \pi_i^* - c_i$, where $i \in [2]$ comprises an optimal solution of the payment-partition scenario.

5.2.1   Equivalence to a Rubinstein Bargaining Game

Initially, we show the equivalence between the *payment-partition* game and a Rubinstein bargaining game, a.k.a. the basic alternating-offers game defined next according to [3]:

Definition 5.2 (Rubinstein Bargaining Game) Assume a game of offers and counter-offers between two players, $\pi_i^r(t)$, where $i \in \{1, 2\}$ and t indicates the time of the offer, for the partition of a cake, of initial size of Π^r. The offers continue until either agreement is reached or disagreement stops the bargaining process.

At the end of each period without agreement, the cake is decreased by a factor of δ_i. If the bargaining procedure *times out*, the payoff to each player is 0. Offers can be made at time slot $t \in \mathcal{N}_0$. If the two players reach an agreement at time $t > 0$, each receives a share $\pi_i^r(t) \cdot t \cdot \delta_i$, where $\delta_i \in [0, 1]$ is a player's discount factor for each negotiation period that passes without agreement being reached. The following equation gives the payment partitions of the two players:

$$\pi_1^r(t) \cdot t \cdot \delta_1 = \Pi^r - \pi_2^r(t) \cdot t \cdot \delta_2. \tag{5.5}$$

So, if agreement is reached in the first negotiation period, the payment partition is as follows:

$$\pi_1^r(t) = \Pi^r - \pi_2^r(t). \tag{5.6}$$

The payoff U_i^r of the players $i, j \in [2]$ if the agreement is reached in iteration t is as follows:

$$U_i^r(t) = \pi_i^r(t) = \Pi^r - \pi_j^r(t). \tag{5.7}$$

Such a game is called a Rubinstein bargaining game.

Proposition 5.1 *Fix a specific quality q. Then, the payment-partition game is equivalent to the Rubinstein bargaining game, when the agreement is reached in the first negotiation period.*

Proof Assuming that an agreement in the Rubinstein bargaining game is reached in the first negotiation period $t = 1$, then the game satisfies the following:

$$U_1^r(1) + U_2^r(1) = \Pi^r(1), \tag{5.8}$$

which is a constant.

In the payment-partition game, assuming an agreement profile s, we have:

$$U_1(s) + U_2(s) = \pi_1 - c_1 + \pi_2 - c_2 = \Pi - c_1 - c_2,$$

since $\Pi = \pi_1 + \pi_2$ and c_1, c_2 are constants for a fixed data protection guarantees level. It follows that $U_1(s) + U_2(s)$ is also constant. It follows that the *Rubinstein bargaining* game and the *payment-partition* game are equivalent.

We define:

Definition 5.3 (Optimal Payment Partition) The optimal partition is when bargaining ends in an agreement profile that gives the highest possible payoff to each player given all possible actions taken by the opponent.

Proposition 5.1 immediately implies:

Corollary 5.1 *Assume that agreement in a Rubinstein bargaining game is reached in the first negotiation period, that $\Pi^r = \Pi$, and that the corresponding profile s^* is an optimal partition for the Rubinstein game. Then, s^* is also an optimal partition for the payment-partition game.*

5.2.2   Designing a Solution to the Payment-Partition Game

Since the Nash bargaining game [3] and the payment-partition game are equivalent, we utilize the solution of a Nash bargaining game in order to compute an optimal solution, i.e. a configuration, which is satisfactory for the two providers in terms of payoffs from the payment-partition game.

The solution of the Nash bargaining game, known as the *Nash bargaining solution*, captures such configuration, where the two bargaining game players are both satisfied. Therefore, since disagreement results in payoffs of 0, we are looking for an agreement profile $s = (\pi_1, \pi_2)$ such that the corresponding partition of the players is an optimal payment partition, i.e. the partition that best satisfies both players' objectives (Definition 5.3). Section 5.3 elaborates on resolving the payment-partition game through the use of the Nash bargaining solution and further addresses issues of truthfulness that arise when applying such configuration.

5.3   Payment Partition Based on the Nash Bargaining Solution

We have already shown how to model cooperation between the HMSP and the CP in order to enable service continuity during a service session for an agreed level of data protection guarantees. Since disagreement is not a desirable strategy for either of the two cooperating players, we may conclude that the two players will reach an agreement. Consequently, the payment must be partitioned in such a way that both the participating providers are satisfied. To reach the optimal solution we utilize the well-known *Nash bargaining solution* [3], which applies to Rubinstein bargaining games when the agreement is reached in the first negotiation period, and therefore to the payment-partition game, given the equivalence presented in Sect. 5.2. Since the Nash bargaining game and the payment-partition game are equivalent, we compute an optimal solution, i.e. a configuration, which is satisfactory for the two providers in terms of payoffs from the payment-partition game. The solution of the Nash bargaining game, known as the *Nash bargaining solution*, captures such configuration. Therefore, since disagreement results in payoffs of 0, we are looking for an agreement profile $s = (\pi_1, \pi_2)$ such that the corresponding partition of the players is an optimal payment partition, i.e. the partition that best satisfies both players' objectives.

The next theorem proves the existence of an optimal partition of the payment between the two players, given each provider's cost c_i.

Theorem 5.1 *There exists an optimal solution for the payment-partition game, and is given by the following:* $\pi_1 = \frac{1}{2}(\Pi + c_1 - c_2)$, $\pi_2 = \frac{1}{2}(\Pi + c_2 - c_1)$.

Proof We consider only agreement profiles and thus refer to the partition π_i assigned to player i. In any such profile it holds that $\pi_1 + \pi_2 = \Pi$. Assuming a disagreement implies that cooperation fails between the two players and the payoff gained by player i equals to $U_i(s^d) = 0$. Since in any such profile, it holds that $U_i(s^a) > 0$ it follows that the disagreement point is not an optimal solution. Since the payment-partition game is equivalent to the Nash bargaining game (Corollary 5.1), a Nash bargaining solution (NBS) of the bargaining game is an optimal solution of the payment-partition game between two players, i.e. a partition (π_1^*, π_2^*) of an amount of goods (such as the payment). According to the NBS properties it holds that:

$$NBS = (U_1(\pi_1^*) - U_1(s^d))(U_2(\pi_2^*) - U_2(s^d))$$

$$= max(U_1(\pi_1) - U_1(s^d))(U_2(\pi_2) - U_2(s^d))$$

$$0 \leq \pi_1 \leq \Pi,$$

$$\pi_2 = \Pi - \pi_1.$$

Since $U_1(s^d) = 0$, $U_2(s^d) = 0$ and $\pi_2 = \Pi - \pi_1$:

$$max(\pi_1 - c_1)(\Pi - \pi_1 - c_2)$$

$$= (-2\pi_1 + \Pi - c_2 + c_1) = 0.$$

Therefore,

$$\pi_1 = \frac{1}{2}(\Pi - c_2 + c_1), \ \pi_2 = \frac{1}{2}(\Pi + c_2 - c_1). \tag{5.9}$$

We proceed to investigate how the solution to the payment-partition behaves when we consider the existence of a constant that represents the ever-existing probability of failure to meet the guarantees level, either due to network service degradation or due to a malicious attack. These events are unlikely but not impossible and are considered as a constant probability considered by the players in this scenario.

Theorem 5.2 *Assume that there exists a constant value p_i^f to provider i representing the expected service degradation or failure, based on the probability of a malicious attack on the infrastructure or the communication link and also based on the current network conditions. Then, the value of the optimal solution is the same as in Theorem 5.1.*

Proof Our game has the same strategy set as before. Concerning the utility functions of the players in case of disagreement, we have also $U_1(s^d) = 0$, $U_2(s^d) = 0$, $\pi_2 = \Pi - \pi_1$ as before. In case of agreement, we have in addition a constant probability p_i^f in the payoff function of each network:

$$U_i(s) = (1 - p_i^f)(\pi_i - c_i). \tag{5.10}$$

Therefore,

$$max(1 - p_1^f)(\pi_1 - c_1)(1 - p_2^f)(\Pi - \pi_1 - c_2)$$
$$= (1 - p_1^f)(1 - p_2^f)(-2\pi_1 + \Pi - c_2 + c_1) = 0.$$

The optimization shows that if constant probabilities are considered, the optimal partition is still as previously, i.e. the providers' cost is the deciding factor for the optimal solution similarly to Eq. (5.9):

$$\pi_1(q) = \frac{1}{2}(\Pi - c_2 + c_1), \ \pi_2 = \frac{1}{2}(\Pi + c_2 - c_1). \tag{5.11}$$

Remark Note that the estimated probability that may be assigned as the probability of service degradation or failure does not affect the optimal partition, but affects each provider's payoff function; therefore, a provider with a high number for such estimated probability of degradation will receive much less of a payoff than the optimal payment partition calculated according to its cost.

5.4 A Bayesian Form of the Payment-Partition Game

Since the partitioning is based on each provider's cost, it is required that the providers are truthful about their costs. Truthfulness is a very important consideration in cooperative situations, especially in bargaining games. The question that arises is whether it would be wise for a player to lie, considering that the player cannot guess whether the other player has more or less cost, thus not being able to correctly assess the risk of such an action. Consider in our scenario that the infrastructure costs of two providers can be easily verified, since often such costs are advertised (within the service advertisements). However, advertised costs are only a part of service costs, which may be different for each provider (Fig. 5.6).

A Bayesian game [1] is a strategic form game with incomplete information attempting to model a player's knowledge of private information, such as privately observed costs, that the other player does not know. Therefore, in a Bayesian game, each player may have several types of behaviour (with a probability of behaving according to one of these types during the game). We use the Bayesian form for the *payment-partition* game, in order to investigate the outcomes of the game, given that each provider does not know whether the cost of its opponent is lower or higher than its own.

Fig. 5.6 In a Bayesian Game each of the interacting players moves sequentially and considers one of various types of behaviours, each of which results in different payoffs for the game

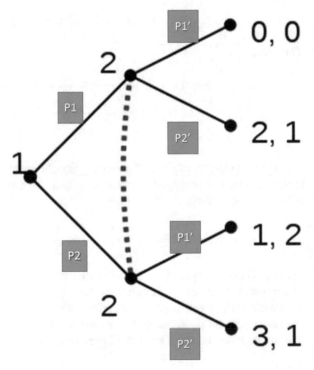

Table 5.2 Provider i payoffs when opponent is of type *lower-cost*

		Provider j actions	
		D	C
Provider i actions			
	D	$\frac{1}{2}(\Pi + c_i - c_j)$	$\frac{1}{2}(\Pi + c_i - c'_j)$
	C	$\frac{1}{2}(\Pi + c'_i - c_j)$	$\frac{1}{2}(\Pi + c'_i - c'_j)$

Table 5.3 Provider i payoffs when opponent is of type *higher-cost*

		Provider j strategies	
		D	C
Provider i strategies			
	D	$\frac{1}{2}(\Pi - c_j + c_i)$	$\frac{1}{2}(\Pi - c'_j + c_i)$
	C	$\frac{1}{2}(\Pi - c_j + c'_i)$	$\frac{1}{2}(\Pi - c'_j + c'_i)$

Let each provider in the *payment-partition* game have two types: the *lower-cost* type (including providers of equal cost) and the *higher-cost* type. Suppose that each of the two providers has incomplete information about the other player, i.e. does not know the other player's type. Furthermore, each of the two providers assigns a probability to each of the opponent's types according to own beliefs and evaluations. Let p_i^l be the probability according to which provider i believes that the opponent is likely to be of type *lower-cost*, and $p_i^h = (1 - p_i^l)$ be the probability according to which provider i believes that the opponent is likely to be of type *higher-cost*.

Since the two players are identical, i.e. they have the same two types and the same choice of two actions, we will only analyse provider i; conclusions also hold for provider j, where $i, j \in [2], i \neq j$. Therefore, provider i believes that provider j is of type *lower-cost* with probability p_i^l, and of type *higher-cost* with probability $1 - p_i^l$. Each provider has a choice between two possible actions: to declare its own real costs (D) or to cheat (C), i.e. declare higher costs $c'_i > c_i$. The possible payoffs for provider 1 are given in Table 5.2 and in Table 5.3.

Lemma 5.1 *If provider i believes that the probability p_i^l, i.e. that provider j is of type* lower-cost, *is higher than the probability p_i^h, then it is more motivated to lie, where $i, j \in [2], i \neq j$.*

Proof In Table 5.2, provider i has higher or equal costs to provider j since provider j is of type *lower-cost*, thus $c_i \geq c_j$. When both players play D, i.e. they both declare their real costs, an equal or greater piece of the payment is assigned to provider i, since the partition of the payment is directly proportional to the providers' costs. If provider i plays C, i.e. cheats, while provider j plays D, then $c'_i > c_i > c_j$, a profitable strategy for provider i, since an even greater piece of the payment will be received. For the cases that provider j decides to play C, then the payment partition may or may not favour provider j (it depends on the actual amount of cheating, and the action of provider i). If provider i plays C, then it is more likely that $c'_i > c'_j$, and provider i will get a greater piece, than if it plays D.

Lemma 5.2 *If provider i believes that the probability p_i^h, i.e. that provider j is of type* higher-cost, *is higher than the probability p_i^l, then it is more motivated to lie, where $i, j \in [2], i \neq j$.*

Proof In Table 5.3, provider i has lower costs compared to provider j, thus $c_i < c_j$. When both players play D, i.e. they both declare their real costs, an equal or greater piece of the payment is assigned to provider j. If provider i plays C, i.e. cheats then $c_i' > c_i$, so playing C will end up in a higher payoff for provider i, and in case provider j plays D, i may even get the bigger piece of the partition. If provider j plays C, it is still better for provider i to play C, since this will end up in provider i receiving a greater piece than it would if it plays D when provider j plays C, although, more likely, not the greater of the two pieces.

5.4.1   Motivating Truthfulness

Proposition 5.2 *Two providers playing the Bayesian form of the* payment-partition *game are not motivated to declare their real costs but instead they are motivated to cheat and declare higher costs, i.e. $c_i' > c_i$, $i \in [2]$, in order to get greater payoffs.*

Proof Straightforward by Lemmas 5.1 and 5.2.

In order to motivate the two providers to declare their real costs, there must exist a mechanism that can penalize a player who turns out to lie on its real cost, assuming that it is detectable whether a player has lied or not. In fact, it is possible to estimate the costs of the two providers considering that some of their costs are advertised and hence known; we refer to such mechanisms as pricing mechanisms [5]. Let the costs be detectable after the service session has terminated, whether either of the participating providers has lied about its costs. In order to motivate the providers to declare their real costs we introduce a *pricing mechanism*, i.e. a new variable that tunes the resulting payoffs, in the payoff function of each player. A side effect for a provider that decides to cheat is that it risks not to be selected for supporting the service in the first place, since by declaring higher costs, the compensation received from the user might be affected, and subsequently, the user might not select the particular provider. The pricing mechanism is a post-game punishment, i.e. cheating in a game does not affect the game in which a provider cheats but subsequent games. Thus, a state of history of a player's behaviour in similar interactions must be kept.

We may define a pricing mechanism consisting of variable $\beta_i \in [0, 1]$, which represents the probability of being truthful, and it may adaptively modify the payoffs of a player. The value of β_i is adjusted at the end of a HMSP-CP interaction, according to the player's behaviour, i.e. whether the player declared its real costs or whether the player lied. We consider that a revelation of the real costs of the two providers is always possible at a later stage of the procedure (e.g. after session termination). The value is adjusted using a *punishment factor* $\gamma \in [0, 1]$ set by a centralized controller affecting the user payment in the future.

Thus, based on $\beta_i^{previous}$, i.e. the previous value of β_i for provider i is defined to be:

$$\beta_i = \begin{cases} \beta_i^{previous} - (\beta_i^{previous} \cdot \gamma), & \text{if provider } i \text{ is caught lying} \\ \beta_i^{previous} - (\beta_i^{previous} \cdot \gamma) + \gamma, & \text{if provider } i \text{ is truthful.} \end{cases} \quad (5.12)$$

Equation (5.12) defines β_i such that on the one hand it decreases fast when cheating behaviour is observed, and on the other hand increases slowly when provider i is truthful, aiming to motivate the players of the bargaining game to remain truthful since the less frequently a player cheats, the closer to 1 its β_i is. The value of $\gamma = \frac{1}{10}$ is reasonable since it allows the faster decreasing and slower increasing behaviour of β_i.

Next, we plot the general behaviour of β_i when $\gamma = \frac{1}{10}$. Thus, Fig. 5.7 illustrates the general form of β_i as it increases from 0 to 1 and decreases back to 0.

The Bayesian form of the payment-partition game is in fact a sequential game, whereas the actual one-shot payment partition is in practice simply a configuration imposed on the two interacting providers. In the Bayesian form of the game, each provider has a memory of its own behaviour during past participations in interactions with other providers in payment-partition games. Thus each provider keeps its own internal state information for its actions, but this does not constitute a history of the game, since the game actions are not directly affected by the internal state maintained by each player. Furthermore, the decision of the opponent player or the history of the game does not depend on this state directly. Indirectly, the action taken by each player at a given interaction may be such as to reflect the information

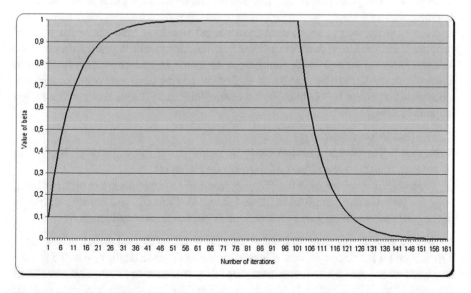

Fig. 5.7 General form of β_i (*increasing and decreasing*)

obtained from this internal state, but this is according to the strategy employed by each provider.

5.5   An Example of Model Evaluation

In this section we evaluate the Bayesian form of the *payment-partition* game between two providers cooperating to support a health monitoring service session requesting data protection guarantees. The payoffs of the game are based on the Nash bargaining solution, as this has been demonstrated in Sect. 5.3, and further include the term β_i, described in the same section, in order to motivate the players to cooperate. The payoff to player i, is

$$\pi_i = \frac{1}{2}(\Pi + \beta_i \cdot c_i - \beta_j \cdot c_j). \tag{5.13}$$

We first run the game (repeated numerical simulations) with the value for β_i always equal to 1, i.e. when no punishment is imposed for lying about costs. The strategies used for the evaluation of the game for Case 1 (no punishment imposed) involve three different strategies for each player, where both providers are allowed to be truthful or lie as follows:

Case 1
Strategy 1   The player always declares the real costs.
Strategy 2   The player always lies about its real costs.
Strategy 3   The player randomly lies 50% of the time and is truthful the rest of the time.

Subsequently, we run another set of numerical simulations but allow the value of β_i to vary, i.e. we allow punishment based on the history of the provider's previous actions. Thus, for Case 2 (with punishment imposed), one more strategy is added to allow the value of β_i to be monitored so that the paying customer could potentially a decision according to its value (or the controlling entity). This additional strategy is the following.

Case 2 (Additional Strategy)
Strategy 4   The player monitors its β_i and only lies if the value of β_i is high, in order to minimize the effects of cheating on its payoff.

The numerical values used for Π, c_i, c_j in case the networks are truthful or lying obey the payoff relations given in Table 5.2 and in Table 5.3 and the overall model of the payment-partition game as described in Sect. 5.2 and resolved in Sect. 5.3.

Specifically, the payment to be partitioned is equal to 10, and the costs vary from 3 to 5, 3 and 4 for the two types of providers when they are truthful, 4 and 5 for the two types of providers when they are lying (they claim costs of one unit more than the actual cost is). Additional work could potentially investigate how the obtained

Table 5.4 Payment-partition
payoffs for player 1, $\beta_1 = 1$

	Provider 2 strategies		
	Only D	Only C	50% C/D
Provider 1 strategies			
Only D	5079.04	4578.54	4829.91
Only C	5579.04	5078.89	5329.42
50% C/D	5329.82	4829.30	5077.09

Table 5.5 Payment-partition
payoffs for player 2, $\beta_2 = 1$

	Provider 2 strategies		
	Only D	Only C	50% C/D
Provider 1 strategies			
Only D	4920.96	5421.46	5170.09
Only C	4420.96	4921.12	4670.59
50% C/D	4670.18	5170.7	4922.91

results would be affected by different cost numbers, e.g. larger costs and unequal lying amounts.

The payoffs vary based on the type of each provider as well as on whether the provider has declared its real costs or has cheated. The provider types are: 1 = *lower-cost* or 2 = *higher-cost* as these are explained in Sect. 5.4. The four strategies that each provider may employ in a simulation run are denoted as follows: (1) always declare its real costs—indicated as *only D*, (2) always cheat—indicated as *only C*, (3) randomly declare real costs or cheat with a probability of 0.5 for each option—indicated as *50% C/D* and (4) only cheat if the value of β_i is high, i.e. cheat only if $\beta_i \geq 0.9$ else declare real costs—indicated as *cheat-if-beta-high*.

First, we present the results for Case 1, when the value of β_i is equal to 1, i.e. when there is no punishment for lying. Tables 5.4 and 5.5 show, each table corresponds to a different player, the cumulative payoffs from the payment partitions that the specific player receives for each strategy combination over 100 simulation runs.

Considering the rows for Player 1, row 2 in Table 5.4 is the highest, showing that the best strategy for Player 1 is *only C*. Also, for Player 2, we consider the columns in Table 5.5. The highest is column 2, thus for Player 2 the best strategy is also *only C*. We observe from these results that the highest payoffs are achieved for each player when he lies and does not declare his real costs, while at the same time his opponent declares the real costs. In any case, for each strategy of the opponent a player's best response is always to lie. Therefore, these results reinforce the theoretical findings of this model, i.e. that if no punishment mechanism is adopted for handling cases when player lies about their real costs, then the players are not motivated to be truthful and declare their real costs.

Next, we present the results for Case 2, when the value of β_i is allowed to vary according to player i's actions. Each player's initial value of β_i in the simulations is randomly generated and then adapted to the player's play according to the selected strategy. This initial value is indicated in the results for reasons of completion.

Table 5.6 Statistics for providers 1 and 2 regarding generation of types

Provider 1 type	Provider 2 type	Number of times
Lower-cost	*Lower-cost*	27
Higher-cost	*Higher-cost*	26
Lower-cost	*Higher-cost*	21
Higher-cost	*Lower-cost*	26

Table 5.7 Average provider 1 payoffs from payment-partition game

	Provider 2 strategies			
	Min. $R = 4$		Max. $R = 1749$	
Avg. $R = 311.67$	Only D	Only C	50% C/D	Cheat-if-beta-high
Provider 1 strategies				
Only D	1570.83	1855.16	1879.71	1591.25
Only C	1265.66	1558.72	1405.64	1301.49
50% C/D	1454.89	1718.82	1566.51	1479.84
Cheat-if-beta-high	1548.79	1827.05	1653.01	1569.23

Table 5.8 Average provider 2 payoffs from payment-partition game

	Provider 2 strategies			
	Min. $R = 4$		Max. $R = 1749$	
Avg. $R = 311.67$	Only D	Only C	50% C/D	Cheat-if-beta-high
Provider 1 strategies				
Only D	1545.84	1251.89	1436.95	1525.41
Only C	1851.01	1557.94	1711.02	1815.18
50% C/D	1661.77	1397.84	1550.16	1638.83
Cheat-if-beta-high	1567.87	1289.62	1463.66	1547.44

Overall, we have run 100 simulation runs with different initial values for β_i and with a randomly generated number of iterations. The number of iterations for each simulation run is indicated in the results as R.

Table 5.6 illustrates some simulation statistics that show how many times in the simulation each network is of either type *lower-cost* or type *higher-cost*, respectively, indicating that there is a uniformity in the generation of the player's types, giving us confidence in the fact that the two networks behave similarly and by examining one of them, we may draw similar conclusions for the other.

Tables 5.7 and 5.8 present the averages of the cumulative payoffs from each simulation run, for all 100 simulation runs. The results are presented for the two providers (provider 1 and provider 2) partitioning the service payment. Tables 5.7 and 5.8 illustrate through the similarity in the related payoffs that the behaviour of the two providers is similar. This is due to the fact that the two providers are motivated by the same incentives and use the same set of strategies. In addition, the generation of provider types in the simulation runs is very similar for the two players, implying that on average the payoffs are eventually very similar for both.

We observe that in all simulation runs, the second strategy, i.e. always to cheat, is the least profitable strategy for both providers, regardless of their types and initial values of β_i. This is because of the presence of β_i in the payoffs, which punishes the choice of cheating, by detecting such an action after any iteration. On the other hand, the other three strategies, which include actions of declaring the real costs, i.e. being truthful, are more profitable strategies. Specifically, the first strategy of always being truthful is the most profitable strategy illustrating how β_i rewards truthfulness, motivating the player to follow strategy *only D*, i.e. always declaring the real costs. In addition, we observe that the fourth strategy of cheating only if β_i is high generates comparable payoffs to the *only D* strategy. This shows that even if the provider decides to employ a strategy, which will allow the provider to cheat a few times much less than the times it cheats when employing the 50% C/D strategy, the best option in terms of payoffs is still to be always truthful.

Furthermore, it is interesting to note that for each provider, while it is more profitable to be truthful, i.e. to declare the real costs, the highest payoffs are accumulated when the opponent decides to cheat, while a provider is truthful. Therefore, if a provider believes that the opponent is more likely to cheat, it is very profitable to always be truthful, i.e. to only declare real costs.

In order to get a better sense of each partition, we provide for each combination of strategies, the percentage partition of the total payment given to the two providers, considering only the amount given and not the amount that would be given if they had cooperated (Table 5.9). We observe that the partitions when the same strategies are used by the two providers are equal or very close to 50% of the amount given, whereas, for the rest of the combinations using the strategy of always declaring the real costs, it results in the greatest partition as it also seen by the corresponding payoffs.

In conclusion, using β_i as a pricing mechanism motivates the interacting providers to declare their real costs, so as to achieve the highest possible payoffs from their interactions.

Table 5.9 Payment partitions for each strategy combination

	Provider 2 strategies			
	Only D	Only C	50% C/D	Cheat-if-beta-high
Provider 1 strategies				
Only D	50.4%, 49.6%	59.7%, 40.3%	56.7%, 43.3%	51.1%, 48.9%
Only C	40.6%, 59.4%	50%, 50%	45.1%, 54.9%	41.8%, 58.2%
50% C/D	46.7%, 53.3%	55.1%, 44.9%	50.2%, 49.8%	47.5%, 52.5%
Cheat-if-beta-high	49.7%, 50.3%	58.6%, 41.4%	53%, 47%	50.3%, 49.7%

5.6   Conclusion

Service and cloud providers may easily need to cooperate in a 5G network due to emerging big data requirements that services may have. In addition, strict data protection requirements may impose additional structures to ensure security and privacy of data, such that a cloud provider is more likely to support than a service provider. Potentially, many different partitions of the service payment may be proposed for the cooperating providers, but since the service and cloud provider must cooperate in order to support the specific health monitoring service customer, the received payment must be partitioned in a satisfying way. Thus, a model of payment partition is proposed such that an equilibrium state of negotiations is reached and both players are motivated to cooperate.

References

1. Gintis, H.: Game Theory Evolving: A Problem-Centered Introduction to Modeling Strategic Interaction. Princeton University Press, Princeton (2000)
2. Harsanyi, J.C.: Games with incomplete information played by Bayesian players. Behav. Sci. **14**, 159–182 (1967)
3. Muthoo, A.: Bargaining Theory with Applications. Cambridge University Press, Cambridge (2002)
4. Nash J.F.: The bargaining problem. Econometrica **18**(2), 155–162 (1950)
5. Nisan, N., Ronen, A.: Algorithmic mechanism design. Games Econom. Behav. **35**(1–2), 166–196 (2001)
6. Rubinstein, A.: Perfect equilibrium in a bargaining model. Econometrica **98**(1), 97–109 (1982)

Index

© Springer Nature Switzerland AG 2020
J. Antoniou, *Game Theory, the Internet of Things and 5G Networks*,
EAI/Springer Innovations in Communication and Computing,
https://doi.org/10.1007/978-3-030-16844-5